FIRE AND EXPLOSION PROTECTION:
A System Approach

ELLIS HORWOOD SERIES IN APPLIED SCIENCE AND INDUSTRIAL TECHNOLOGY

Series Editor: Dr D. H. SHARP, OBE, former General Secretary, Society of Chemical Industry; formerly General Secretary, Institution of Chemical Engineers; and former Technical Director, Confederation of British Industry.

This collection of books is designed to meet the needs of technologists already working in fields to be covered, and for those new to the industries concerned. The series comprises valuable works of reference for scientists and engineers in many fields, with special usefulness to technologists and entrepreneurs in developing countries.

Students of chemical engineering, industrial and applied chemistry, and related fields, will also find these books of great use, with their emphasis on the practical technology as well as theory. The authors are highly qualified chemical engineers and industrial chemists with extensive experience, who write with the authority gained from their years in industry.

Published and in active publication

PRACTICAL USES OF DIAMONDS
A. BAKON, Research Centre of Geological Technique, Warsaw, and A. SZYMANSKI, Institute of Electronic Materials Technology, Warsaw
POTTERY SCIENCE: Materials, Processes and Products
A. DINSDALE, lately Director of Research, British Ceramic Research Association
MATCHMAKING: Science, Technology and Manufacture
C. A. FINCH, Managing Director, Pentafin Associates, Chemical, Technical and Media Consultants, Stoke Mandeville, and S. RAMACHANDRAN, Senior Consultant, United Nations Industrial Development Organisation for the Match Industry
OFFSHORE PETROLEUM TECHNOLOGY AND DRILLING EQUIPMENT
R. HOSIE, formerly of Robert Gordon's Institute of Technology, Aberdeen
MEASURING COLOUR
R. W. G. HUNT, Visiting Professor, The City University, London
MODERN APPLIED ENERGY CONSERVATION
Editor: K. JACQUES, University of Stirling, Scotland
PAINT AND SURFACE COATINGS: Theory and Practice
Editor: R. LAMBOURNE, Technical Manager, INDCOLLAG (Industrial Colloid Advisory Group), Department of Physical Chemistry, University of Bristol
CROP PROTECTION CHEMICALS
B. G. LEVER, Development Manager, ICI plc Plant Protection Division
HANDBOOK OF MATERIALS HANDLING
Translated by R. G. T. LINDKVIST, MTG, Translation Editor: R. ROBINSON, Editor, Materials Handling News. Technical Editor: G. LUNDESJO, Rolatruc Limited
FERTILIZER TECHNOLOGY
G. C. LOWRISON, lately of Fisons Fertilisers Ltd.
NON-WOVEN BONDED FABRICS
Editor: J. LUNENSCHLOSS, Institute of Textile Technology of the Rhenish-Westphalian Technical University, Aachen, and W. ALBRECHT, Wuppertal
MECHANICS OF WOOL STRUCTURES
R. POSTLE, University of New South Wales, Sydney, Australia, G. A. CARNABY, Wool Research Organization of New Zealand, Lincoln, New Zealand, and S. de JONG, CSIRO, New South Wales, Australia
PROFIT BY QUALITY: The Essentials of Industrial Survival
P. W. MOIR, Consultant, West Sussex
MICROCOMPUTERS IN THE PROCESS INDUSTRY
E. R. ROBINSON, Head of Chemical Engineering, North East London Polytechnic
SINGLE CELL PROTEIN: Research, Development, Manufacture
D. H. SHARP, OBE, former General Secretary, Society of Chemical Industry; formerly General Secretary, Institution of Chemical Engineers; and former Technical Director, Confederation of British Industry
QUALITY ASSURANCE: The Route to Efficiency and Competitiveness, Second Edition
L. STEBBING, Quality Management Consultant
INDUSTRIAL CHEMISTRY: Volumes 1 and 2
E. STOCCHI, Milan, with additions by K. A. K. LOTT and E. L. SHORT, Brunel
REFRACTORIES TECHNOLOGY
C. STOREY, Consultant, Durham, former General Manager, Refractories, British Steel Corporation
COATINGS AND SURFACE TREATMENT FOR CORROSION AND WEAR RESISTANCE
K. N. STRAFFORD and P. K. DATTA, School of Material Engineering, Newcastle upon Tyne Polytechnic, and C. G. GOOGAN, Global Corrosion Consultants Limited, Telford.
MODERN BATTERY TECHNOLOGY
C. D. S. TUCK, Alcan International Ltd., Oxon
FIRE AND EXPLOSION PROTECTION: A system Approach
D. TUHTAR, Institute of Fire and Explosion Protection, Yugoslavia
PERFUMERY TECHNOLOGY 2nd Edition
F. V. WELLS, Consultant Perfumer and former Editor of *Soap, Perfumery and Cosmetics,* and M. BILLOT, former Chief Perfumer to Houbigant-Cheramy, Paris, Président d'Honneur de la Société Technique des Parfumeurs de la France
THE MANUFACTURE OF SOAPS, OTHER DETERGENTS AND GLYCERINE
E. WOOLLATT, Consultant, formerly Unilever plc

FIRE AND EXPLOSION PROTECTION: A System Approach

DINKO TUHTAR
Institute of Fire and Explosion Protection,
Senior Scientific Adviser
Sarajevo, Yugoslavia

ELLIS HORWOOD LIMITED
Publishers · Chichester

Halsted Press: a division of
JOHN WILEY & SONS
New York · Chichester · Brisbane · Toronto

First published in 1989
ELLIS HORWOOD LIMITED
Market Cross House, Cooper Street,
Chichester, West Sussex, PO19 1EB, England
The publisher's colophon is reproduced from James Gillison's drawing of the ancient Market Cross, Chichester.

Distributors:
Australia and New Zealand:
JACARANDA WILEY LIMITED
GPO Box 859, Brisbane, Queensland 4001, Australia
Canada:
JOHN WILEY & SONS CANADA LIMITED
22 Worcester Road, Rexdale, Ontario, Canada
Europe and Africa:
JOHN WILEY & SONS LIMITED
Baffins Lane, Chichester, West Sussex, England
North and South America and the rest of the world:
Halsted Press: a division of
JOHN WILEY & SONS
605 Third Avenue, New York, NY 10158, USA
South-East Asia
JOHN WILEY & SONS (SEA) PTE LIMITED
37 Jalan Pemimpin # 05–04
Block B, Union Industrial Building, Singapore 2057
Indian Subcontinent
WILEY EASTERN LIMITED
4835/24 Ansari Road
Daryaganj, New Delhi 110002, India

British Library Cataloguing in Publication Data
Tuhtar, D. (Dinko)
Fire and explosion protection.
1. Industries. Fire protection. Manuals
2. Industries. Explosions. Safety aspects
I. Title
628.9′22

Library of Congress Card No. 88–38306

ISBN 0–7458–0502–7 (Ellis Horwood Limited)
ISBN 0–470–21340–X (Halsted Press)

Typeset in Times by Ellis Horwood Limited
Printed in Great Britain by Hartnolls, Bodmin

Contents

Preface

Although the field of fire and explosion protection is relatively a large one, it is still seldom studied *per se* owing to the complexity of the phenomena which make up sources of fire and explosion hazards and their control. The multidisciplinary nature of fires and explosions has long deterred a coherent scientific approach to this problem, and only recently has the fire and explosion sphere of activity emerged as a distinct scientific discipline. Consequently, the literature strictly concerning the field of fire and explosion protection has been rather scarce. Of course, there are many good works dealing with certain aspects of the fire and explosion problem, for example, physical and chemical fundamentals of combustion, heat transfer, fluid mechanics, etc., but they rarely address the totality of fire and explosion protection.

The idea for this book emerged from my experience as a scientific adviser in the field of fire and explosion protection. In writing this book, two different paths were possible. One was to write a rigorous theoretical treatment of certain areas required for understanding the problem of fire and explosion protection, and the other one was to employ a purely descriptive method. However, both parts seemed inappropriate. While the first one may be of interest only to a relatively small fire and explosion community (and there are a few good books written in this way), the other one has already been covered with numerous popular booklets, brochures, and similar educational and propaganda material. Furthermore, very few works have treated fire and explosion protection as an integrated system composed of an interdependent set of components. Therefore, I have opted for a middle approach, that is to write a book which would contain quantitative treatment only in amounts necessary for understanding the essentials, while retaining a descriptive and system approach more suited for the nonscientist and the beginner. Fire experience reveals that the latter category is precisely the one which, through negligence and lack of elementary knowledge of prevention, causes the greatest number of fires and explosions.

Since fires and explosions are essentially chemical phenomena, the book has been written from a chemical point of view, reflecting the author's own background in chemistry.

Writing a book involves long hours of loneliness. I am indebted to my wife Dinka and little daughter Matea who have been missing me for the last year and a half.

Dinko Tuhtar

1

Fire and explosion protection as a system

1.1 INTRODUCTION

Human interest in fire safety probably dates from the discovery and employment of fire. The obvious benefits of numerous friendly uses of heat energy as one of the main products of combustion are often overshadowed by the enormous destructive power of fires and explosions. Hence there has been continuous interest in understanding the causes of such perils and in devising means of their elimination or reduction.

The first attempts were addressed, logically, to the extinguishment of fires once started, with water being a sole extinguishing agent for a long time. Other agents for putting out fires have been discovered rather recently, and in comparison with water even today are used less frequently. Fire fighting techniques and tactics have progressed to such a stage that at the present time most fires, and even some explosions, can successfully be fought.

On the other hand, fire and explosion prevention, as a narrower aspect of fire protection, except for some rudimentary elements, has lagged behind in its development and application. Most people rarely perceive fire as a significant personal threat, and think of it only in relation to others. Of course the same may be said of other classes of hazards (illness, environmental and occupational risks). In addition, the inherently highly interdisciplinary nature of prevention measures has long deterred the emergence of coherent and logically consistent protection system. While much knowledge on many aspects of hazard protection has accumulated, an integrated, logically sound system has long been lacking. Yet, the best results in achieving as high a degree of fire and explosion safety as possible would be obtained by considering all the elements of fire and explosion protection as an integrated system, consisting of an interdependent set of components.

1.2 EXISTING APPROACHES

There have been only a few attempts at considering fire protection as a system, and usually even these refer only to a specific problem of fire safety, rather than addressing fire protection in all its totality.

Smith (1983) pointed out that for an effective hazard prevention in industry it is necessary to analyze all the conditions that could give rise to risk factors, and on this basis work out the best protection measures. An effective linkage of organizational and technical measures is a must.

Hoffman (1983) described the principles on which industrial fire safety is based in the Federal Republic of Germany. A large number of construction and architectural measures, coupled with numerous fire prevention requirements, technological and production characteristics and administrative control, should be adequate to accomplish this purpose.

Cox & Pasman (1983), referring to the risk analysis in the process industries, commented on the different methods of risk analysis which lack both the necessary coherence and the common terminology. However, each risk analysis, in their opinion, should be able to answer the following questions. What may go wrong? What possible effects may result? How frequently will a malfunction or disaster occur? Can it be minimized and in what way? A complete risk analysis is very costly and takes a long time.

Lowe & Solomon (1983) also presented recommendations for hazard identification procedures in the process industries. The effectiveness of hazard identification depends on a number of factors, among which are the availability of information on process technology and characteristics of materials employed in the process, the quality of the existing control measures and the emergency procedures in case of accidents. Since it is impossible to develop a universally applicable system of hazard identification, it is necessary for each process to plan a specific procedure.

Marshall (1980) reviewed historical and theoretical approaches to the prediction of hazard and risk in general. He defined hazard as a physical situation with a potential for human injury, while risk may be defined as the measure of the frequency of the realisation of a hazard. Hazard does not have dimensional representation, whereas risk can be expressed as frequency with dimensions of reciprocal time. The methodology for predicting the risk of failure of a system is based on the construction of a 'fault tree' containing many 'and' and 'or' gates for the identification of hazardous events.

Kunzi (1980) proposed a systematic program for investigations of thermic hazards in the process industries. The 11 step program is based on a number of 'if yes' and 'if no' decisions, while the practical procedure is carried out by using a separate scheme. Although such a pragmatic approach to the problem of thermic safety proved to be satisfactory for a routine investigation, some uneasiness as to the correct decision at some stages of the program may be felt by the examiner.

In assessing hazards of combustion products in building fires, Fowell (1984) described a framework whose three chief components are the generation of combustion products, the movement of those products around the building and the response of the occupants. He discussed current measurement and modeling capabilities of each of these three subcomponents of toxic hazard assessment, and noted that the refinement and the completion of the framework will take many years.

1.3 NEWER CONCEPTS

The most comprehensive qualitative approach to a system concept of fire safety in buildings today seems to be the one developed by the NFPA Committee on System

Concepts for Fire Protection in Structures (Connelly & Harvey 1986). It urges that building fire safety should be incorporated in building design as an integrated subsystem on the same terms as other necessary subsystems (aesthetical, functional, structural, electrical and mechanical). The organizational scheme of the system is based on a decision tree, employing the logic and structure of a fault tree, which describes the elements and their interrelationship that must be considered in building fire safety. The tree itself is rather branchy, and is somewhat cumbersome for practical use because of its excessive size. Nevertheless, the logical sequence of various events that must be followed in achieving previously identified fire safety objectives ensures a high degree of fire protection in structures early in the building design stage. The structure of the tree, its applications, and limitations are detailed in an NFPA guide (NFPA 550 1986).

At the same time some quantitative approaches have evolved. These try to quantify particular components of fire protection by assigning to them corresponding probabilities or certain numerical values. The Swiss Insurance Method is often used in Europe for the purpose of property protection (Kaiser 1979/80). The fire risk of a facility is determined as a ratio of hazard and protection factors. The fuel load of a structure, the combustability of materials, the potential for smoke generation, etc., determine the hazard factor, while the protection factor depends on the existing fire protection systems.

The identification and ranking of inherent hazards of a plant using only pre-design stage data are made possible by using the Mond Index (Tyler 1985). The plant is divided into a series of units which are separately assessed. For each unit a numerical index is calculated and converted to a qualitative description of the process (from low to very high). If a particular product can be made by different routes, the Mond index enables one to select the more inherently safe route.

Fire protection can be depicted in the form of network diagrams (Connelly & Harvey 1986). A major prerequisite is a clear recognition of all the events that are needed to accomplish the stated objective and the inclusion of the other alternatives in reaching that objective. The probability values for each event are determined by using multiplication and addition as the only mathematical operations. In this way a statistical weight factor could be obtained for two or more alternative fire protection designs.

A method that addresses fire protection of some mass assembly occupancies such as health care, detention and correctional occupancies has been developed by the National Bureau of Standards in the USA. This Fire Safety Evaluation System (Connelly and Harvey 1986) calculates the risk level of an occupancy on the basis of previously assigned numerical values for each of the controlling safety factors involved. The system is currently being developed for other types of occupancies with the hope that it will be adopted for all buildings.

Finally, some infinite state transition models have recently been developed and used for fire protection. Details are given elsewhere (Connelly & Harvey 1986).

1.4 THIS BOOK'S APPROACH

The concept of treating fire protection as a system is quite new. So far it has only been applied, as has just been explained, to some rather limited segments of overall fire protection. Moreover, the field of explosion protection has not yet been addressed explicitly as a logical 'subsystem' within the system of fire protection. A condensed, simple but encompassing system of fire and explosion protection, based on the already accumulated knowledge on fire protection, is lacking.

This book describes the principles of such a system. The approach used is not new, however. A similar concept of the system approach was used earlier by (Seinfeld 1975) in the field of air pollution protection. Tuhtar (1984) extended Seinfeld's methodology to water protection, and showed that it could be applied to general environmental protection.

In the field of environmental protection it is the hazard that a pollutant may bring that is of most concern. Its effects on man and his environment are such that an effective strategy for the control of emission should be sought. Control can be accomplished by employing both technical and legislative measures which, if enforced strictly, should ensure that the negative effects of pollutants be either absent or reduced to some acceptable level. Thus, the effects of pollutants, the sources of emission, and the means of their control form a framework for a system approach to environmental protection (Seinfeld 1975, Tuhtar 1984).

Close examination of such a system reveals that it can also be successfully applied to the problem of fire and explosion protection. The principal difference lies really only in the nature of the hazard, with increased temperature, or heat energy, being the 'pollutant' in the fire protection system. A schematic presentation of this system as applied to fire and explosion protection is shown in Fig. 1.1. Fig. 1.1 shows the

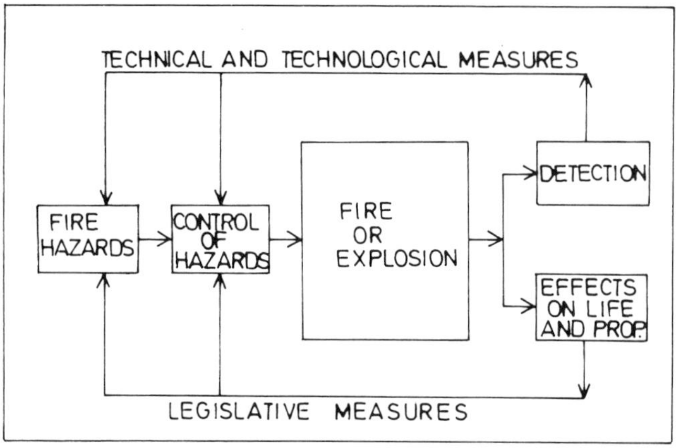

Fig. 1.1 — Fire and explosion protection as a system.

components of the fire and explosion protection system and their interrelationship.

The principal objectives of the system, i.e. life safety and property protection, can be achieved only if all existing and potential fire and/or explosion hazards have been identified and characterized. The chemical reaction between a combustible material and an oxidant, aided by a heat energy source as an initiator of a fire or explosion, serves as the starting point for hazard identification. Numerous combustible materials can be conveniently classified according to their aggregate state as solids, liquids, gases and dusts. Fire and explosion characterization of these materials, using physicochemical methods of analysis, enables the most hazardous ones to be pinpointed. Oxygen from air is in most cases the only oxidant reacting with combustibles, which means that as a hazard it may be considered as somewhat constant, owing to its overwhelming presence. This leaves sources of heat energy of sufficient intensity to start the combustion/explosion reaction as the next most important factor that is of principal interest in defining fire and explosion hazards. This component of the fire and explosion system is dealt with in Chapter 2.

Once the fire and explosion hazards have been identified, the next logical step in the protection system is to find a means of controlling the hazards. Although there are many methods for the reduction of fire and explosion hazards, they really belong to one of two broad classes: preventive measures, which are undertaken before a fire or explosion is initiated, and repressive measures which are employed after. Moreover, it will be shown that all fire and explosion protection measures are based on one simple principle, that is the exclusion of at least one of the reactants in the chemical reaction causing a fire or explosion. Even though all the components of a fire and explosion protection system are indispensable and mutually dependent, the component that covers the control of fire and explosion hazards is by its nature the heart of the protection system. Most of the materials in the book refer to the means of achieving the needed protection against fire and/or explosion. Thus, Chapter 3 describes the essentials of fire and explosion hazard control, while Chapters 7 and 8 detail the most important technical and legislative protection measures.

Fires and explosions, being probabilistic events, can never be entirely prevented, regardless of the quality of measures for control of hazard sources. For protection purposes the most important are the very early stages of a fire and explosion, as later stages usually cause substantial material and often human losses. The dynamics of fire and explosion are reviewed in Chapter 4.

In order to reach the stated objectives of fire safety it is necessary to detect a fire or explosion at the earliest possible time. Detection can be achieved by employing either instrumental detection (Chapter 5), or, in principle, some of the effects of fire and explosion on man and his environment (Chapter 6). These two components of the protection system form a basis for the use of a number of technical measures (Chapter 7) which act upon the fire or explosion hazards or the control of them in such a way as to prevent, or, if the fire has already been initiated, to extinguish it. The technical measures are implemented by corresponding legislative measures (Chapter 8) which are aimed at the same components of the system, i.e. the sources of hazards and their control.

The proposed system of fire and explosion protection is general in nature. So are the solutions based on it. It can serve as a qualitative guide only. However, it can easily be applied to concrete fire and explosion hazard problems, taking into consideration the specifics and uniqueness of each problem. This capability of satisfying multiple fire and explosion protection objectives distinguishes it from the other existing system approaches. Although the system does not contain time factors — nor does it provide a quantification measure as a final output — its logical sequence of steps to be followed in fire and explosion protection problem solving does enable a systematic analysis of ways to reach the previously defined objective(s) to be achieved.

2

Sources of fire and explosion hazards

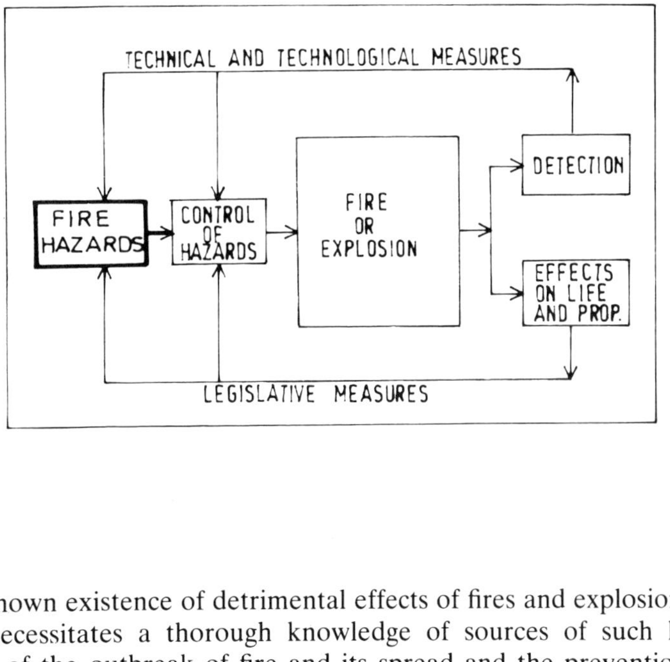

The well known existence of detrimental effects of fires and explosions on life and property necessitates a thorough knowledge of sources of such hazards. The prevention of the outbreak of fire and its spread and the prevention of fire and explosion losses are possible only if the causes of fire and explosion are precisely determined. This implies that answers to the questions such as why a fire started, why it spread (if it did), and why fire losses occurred should be known. Much of the problem can be solved if one ascertains what were the materials involved, and what were the nature and intensity of the source of ignition energy.

The most frequent way of finding answers to the above questions includes intelligent use of the much publicized 'fire triangle' form, or of the recent 'tetrahedron' form for the flaming mode of a combustion process (Haessler 1986).

However, a closer examination would reveal that both of them could really be referred to the simple fire and/or explosion chemical equations (2.1) and (2.2) which state that for a fire or explosion to occur, three basic requirements must be fulfilled: (1) there must exist a flammable or combustible material, which acts as a chemical reductant (often referred to as a fuel), (2) there must exist a source of oxygen (most often air), or some other chemical oxidant which is capable of oxidizing the reductant, and (3) a source of heat energy and a means of bringing this energy into contact with the reductant must also exist (shown as the small triangles in equations (2.1) and (2.2).

$$\boxed{\text{Reductant}} + \boxed{\text{Oxidant}} \xrightarrow[\Delta]{} \boxed{\text{Products}} \quad (2.1)$$

$$\boxed{\text{Fuel}} + \boxed{\text{Air}} \xrightarrow[\Delta]{} \boxed{\text{FIRE or EXPLOSION}} \quad (2.2)$$

The discussion that follows will cover in some detail the most relevant facts from a fire and/or explosion point of view regarding each of the principal blocks of the above equations.

2.1 HAZARDS OF MATERIALS

Most materials which surround man are combustible, that is capable of a vigorous chemical oxidation at elevated temperatures. The few classes of incombustible materials are mostly of inorganic origin. Fire and explosion hazards of a material are markedly dependent on its aggregate state. Flammable gases and vapours are by far the most hazardous, followed by flammable and combustible liquids, and some solids in the form of fine dust. Ordinary solids are not that hazardous, except in some exceptional circumstances (see later in this section). Therefore, it seems convenient to discuss fire and explosion hazards of materials according to their aggregate state. Explosives are a special class of hazard and are usually treated separately (see section 2.4).

2.1.1 Solids

2.1.1.1 Fire characteristics

The most significant characteristics of solids that determine their fire and explosion hazard potential are the following.

Combustibility

This parameter is one of the basic criteria for the assessment of fire hazards of solids, as it gives an unambiguous answer as to the capability of a material to burn. It is widely used for the purpose of evaluating the usefulness of a material to be used in building construction. Materials that are combustible, if unprotected, are normally ruled out as suitable for construction.

The essence of the test method for the evaluation of combustibility of a solid material is through the introduction of a sample specimen into a furnace previously heated to a predetermined temperature (normally between 1023 and 1173 K), and in observing after some short time (a few minutes) the behavior of the sample in the furnace and when taken outside (ISO 1182/1983). Depending on its capability for an unsustained burning, the material can be broadly classified as either combustible or noncombustible. If combustible, gradations such as 'easily', 'moderately', or 'combustible with difficulty' are possible, contingent on the intensity and duration of contact with the ignition source.

The combustibility of a material, not being a fundamental characteristic, depends on a number of variables, such as temperature, pressure, volume, size of particles, etc.

Since the standard method for combustibility testing used with materials which are on the borderline of combustibility frequently gives nonreproducible or contradictory results, a modification of this method, based on the application of oxygen-consumption calorimetry, has been proposed. Instead of measuring temperatures and using other indexes of test results, the suggested modification offers the quantity of heat liberated from the sample as a small reliable value (Ahonen *et al.* 1985).

The majority of combustible solids are of organic origin.

Materials which have been found to be noncombustible can still be hazardous since at higher temperatures they may emit dense smoke and toxic gases (Hertel 1985).

Spread of flame

Knowledge of the speed of flame propagation over solid materials is important for life safety considerations. Materials that readily spread flame are unsuitable for a wider application, particularly in public buildings, hospitals, schools, museums, etc. There are many tests for the spread of flame. the best known and widely used are the Steiner Tunnel Test, the Radiant Panel Test, and the Corner Test (DeHaan 1986).

The rate of flame spread of a material is a function not only of its chemical and physical characteristics, but also of its orientation in the environment, the intensity of the incident radiation, temperature and the rate of air movement, the thickness and width of the sample, the composition of the surrounding atmosphere, and the extent of fire. The altitude at which the determination is performed (that is, the atmospheric pressure) is also of some significance (Starrett 1981) since it was found that the rate of flame spread decreases with decreasing atmospheric pressure. It is generally accepted that geometrical factors, such as the surface area and the orientation of the sample, have a much greater bearing on the rate of flame spread over a solid material than do its chemical and physical characteristics.

Unfortunately, data on the rate of flame spread of a material obtained in laboratories can seldom be well correlated with the behavior of the same material in real fire scenarios. Materials which spread flame rather slowly under small-scale laboratory testing conditions often behave quite differently in real fires, or vice versa, owing to differing mechanisms of flame spread. While the rate of flame spread over a solid surface under laboratory conditions is determined primarily by factors such as the local chemical kinetics, the heat conductivity and gas diffusion, the controlling factors during a real fire are the duration and intensity of heat release, as well as the intensity of heat feedback from the flame to the surface of the burning material (Ris 1985).

Quintiere & Harkleroad (1985) developed an experimental procedure which can be used to obtain data for the theoretical prediction of ignition and flame spread of materials in natural convection in opposed air flow by using mathematical fire models. The relevant data are the thermal absorptivity of the material $(K\rho C)^{1/2}$ (see section 4.2), its ignition temperature, and a parameter θ related to the available flame energy.

Rate of heat release

Data on the rate of heat release are among the most important fire characteristics of solid materials. It has been found that laboratory test methods which measure the rate of heat release with a cone calorimeter can be successfully used for the prediction of the amount of heat released during the combustion of materials in a structure (*Int. Fire Chief* 1987, Babrauskas 1986a, 1984). The principle of the method is based on the fact that for different organic materials the heat produced per equivalent of oxygen consumed is approximately constant and equals 13.1 (+/− 10%) MJ/kg O_2. Thus, by measuring the oxygen content at the inlet and outlet of the cone calorimeter, the amount of oxygen consumed can be determined and the heat released by the burning material at various stages of combustion can be calculated. The apparatus, developed at the US National Bureau of Standards, also enables measurements of some other fire characteristics of solid materials to be taken such as the formation of soot, smoke, and the behavior of materials under the influence of radiation sources of differing intensity.

Smoke production

Smoke produced in fires in enclosures presents one of the principal hazards to life safety, property protection and fire suppression. Most fires are accompanied by smoke production throughout all the phases of burning, that is ignition, growth, and decay. Fire statistics reveal that the majority of victims in fires come from the inhalation of smoke and toxic gases and from the serious reduction in visibility which sometimes makes an orderly evacuation of people from buildings impossible (see subsection 6.2.1). Thus it is important to have some information on the smoke-producing capacity of solids in fires. These data are needed for the design of paths of egress and smoke control systems.

Very large amounts of very dense, sooty, black smoke are liberated during the burning of some plastics. Lowered flammability of some plastics achieved by fire retardant chemicals is sometimes more than offset by increased smoke production.

Smoke generation from a solid material is dependent on a number of factors, such as the chemical characteristics, the additive present, the mode of burning (whether flaming or smoldering) and the rate of heat release. The geometry of the burning material and the intensity of air ventilation significantly influence smoke production (Egan & Litton 1986). The air ventilation is especially important. Smoke production for the same material can be changed four to five fold, depending on the ventilation (Drysdale & Abdul-Rahim 1985).

The rate of smoke production from a fire in an enclosure can be estimated from a simple expression (equation (2.3)) (Butcher & Parnell 1979):

$$m = 0.188\, P\, y^{3/2} \tag{2.3}$$

where m=the mass rate of smoke production (kg/s), P=the perimeter of the fire (m) and y=the distance from the floor to the bottom of the smoke layer (m).

Since dense smoke can seriously lower the visibility it is instructive to consider the relationship between these two factors. A quantitative measure of smoke density can be obtained by determining the optical density of smoke, OD_x, defined as:

$$OD_x = \log \frac{I_o}{I_x} \tag{2.4}$$

where I_o and I_x are the intensities of the incident and transmitted light passing through a smoky enclosure of x m. It has been shown that the Lambert–Beer law in analytical chemistry, which relates the optical density to the concentration of the absorbing solution, can be used in the case of smoke as well. The visibility in an enclosure filled with smoke is inversely proportional to the optical density per meter (equation (2.5) and Fig. 2.1):

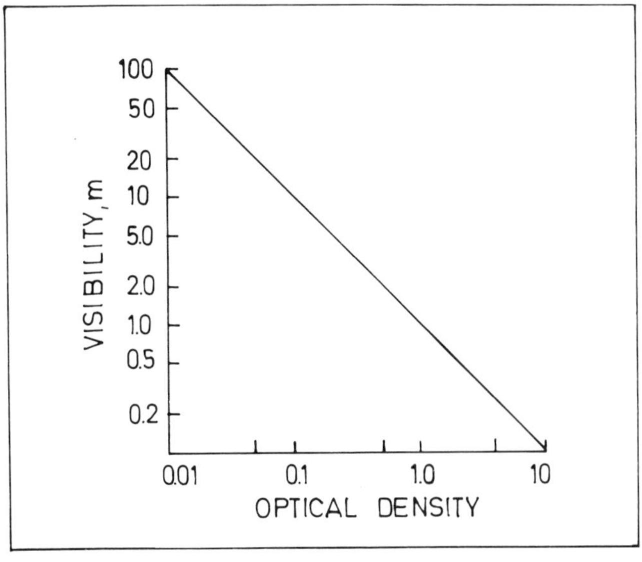

Fig. 2.1 — Visibility in a smoky enclosure as a function of optical density of smoke.

$$\text{Visibility} = 1/\text{OD} \tag{2.5}$$

It can be seen from Fig. 2.1 that an increase in optical density from, say, 0.1 to 2.0 leads to a 20-fold reduction in the visibility (from 10 m to just 0.5 m).

The existing test methods for determining the smoke generation of solid materials are still used mainly in research development and production quality control of materials and products. The extent to which smoke measurements are reliable for assessing the smoke hazard in actual fires is still unknown. However, it is felt that results of test methods may be informative enough for a rough analysis of smoke production during an actual fire (Christian & Nolan 1985). In fact, experimental data and theoretical calculations have shown at least that materials with a pronounced propensity for smoke generation under the conditions of large-scale fire tests also generate a lot of smoke in laboratory tests (Paul 1985).

The obvious goal of smoke testing is to rank materials according to their tendency for smoke development in actual fires. The test method closest to the theoretical ranking of materials is the ISO method DTR 5924 (project).

Toxicity

Irrespective of the kind of materials subjected to burning, the products of combustion are always toxic to a lesser or greater degree. Even in the case of the ideal combustion of a material producing just H_2O and CO_2 as the final and only products, the danger to people would come from asphyxiation due to the replacement of oxygen in the air by CO_2. Since ordinary combustion reactions are not ideal, and sometimes are far from it, there is always the production of a small amount of partially oxidized material, in the form of smoke and gases, notably carbon monoxide, CO. This invisible, tasteless, and odorless gas is not the most toxic of the many gaseous products liberated in fires, but is the most abundant of all the toxic gases (up to a few volume percent). Its toxicity has been known for a long time and is primarily due to the formation of a strong complex, carboxyhemoglobin, in the blood, with the result that the body tissues are deprived of needed oxygen.

With the advent of new materials, such as different kinds of plastics, of which there are thousands in number, concern has been expressed as to the possible toxic effects of gases liberated in fires of such materials. To improve different physical characteristics of plastics during the synthesis and polymerization, small amounts of other elements are added either in the form of monomer, or as an additive. The resulting gaseous products of combustion contain, in addition to CO, small amounts of highly toxic gases such as nitrogen oxides, NO_x (x=1,2), sulfur oxides SO_x (x=2,3), hydrogen cyanide, HCN, hydrogen halogenides, HX(X=F^-,Cl^-,Br^-), etc. Chemical analysis of the atomsphere in fire experiments with the burning of wood and some solid plastics in a full-sized enclosure (3.3×3×2.4 m), having one opening connected to a corridor of size 1.17×2.4×11.4 m, disclosed the presence of 20–40 different species (Fardell *et al.* 1986). Besides CO_2 (10–12% vol.%) and CO (1.2–4.0% vol.%) as the principal fire gases, many organic volatiles, consisting of aromatic and partially oxidized hydrocarbons in ppm concentrations, have been found. The average molecular weight of organic volatiles was 142. It has been concluded that the composition of the fire atmosphere is much more dependent on the intensity of fire and temperature in the fire centre than on the type of material being burned.

As the use of plastic based materials and products has reached phenomenal proportions, one would expect an increasing trend in the number of fatalities from fires. However, this has not been so, since the number of fire fatalities in many countries is actually decreasing. This implies that the toxicity of many plastics does not differ much from the toxicity of wood, cellulosics, coal and other ordinary solids.

Results of many toxicity tests with plastics seem to support this conclusion (Braun & Levin 1986).

Firemen and people trapped in fires, if unprotected, experience the direct effects of smoke and fire gases. The effects can range from relatively light headaches, to nausea, to unconsciousness, and to death. This form of toxicity, termed acute toxicity, is difficult to measure using people as experimental objects. Instead, the response of experimental animals, usually rodents, exposed over a certain time to different concentrations of a combustion atmosphere is determined. A concentration/response relationship, commonly in the form of a straight line, is obtained, from which the concentration producing a response effect in 50% of the tested animals within the specified time, termed LC_{50}, can be calculated. Data obtained using animals are not easily extrapolated to people, though. The expertise and judgment of a qualified toxicologist are needed before conclusions from LC_{50} values can be drawn. In addition, the question of validity of correlation of laboratory tests with real fire scenarios further complicates the aim of finding acute toxicity data for people.

Research on the toxicity of fire gases is still expanding. The goal is to find a means of regulating the toxicity of materials in buildings. A fast method for the evaluation of the toxicity of gases emitted in fires of different materials has been proposed (Reichwerger & Balbin 1985). It offers a means of classifying the materials according to the toxicity of their fire gases compared with the toxicity of wood (oak) as the reference material. Mass concentrations of six principal toxicants (CO, HCN, NO_x, HCl, SO_2 and aldehyde) have been determined colorimetrically, using Dräger tubes. Samples of different materials have been ignited in a furnace kept 50 K below and above the ignition point. The toxicity of the gases has been expressed in the form of a toxicity coefficient, R, defined as:

$$R = \frac{n_i}{t_i} \tag{2.6}$$

where n_i is the maximum concentration of the ith gas, and t_i is the limit of toxicity of the ith gas during a 10 minute exposure. The values of t_i for CO, HCN, NO_2, HCl, SO_2 and aldehyde have been set as: 9000, 350, 200, 500, 500 and 65 ppm, respectively. Knowing the volume of the experimental enclosure, V, and of the sample, a, a toxicity index, I, can be defined as:

$$I = (V/a)R \tag{2.7}$$

A toxicity classification of materials has been proposed according to the values of the toxicity coefficient, R. Materials with a value of $R > 100$ are designated as very dangerous, since their toxicity is considerably higher than that of wood. Materials with an R value $10.1 < R \leq 100$ are approximately equally toxic as wood, while materials with a value of R $1.1 < R < 10$ are less toxic than wood. Finally, an R value below 1 indicates a nontoxic material.

Whether the proposed classification will be widely accepted as a means of regulating the toxicity of materials in buildings remains to be seen.

Heat of combustion

The heat of combustion of a solid material, defined as the amount of heat produced when it burns, has long been one of the primary indicators of its fire hazard. The synonyms of this characteristic are the *enthalpy of combustion,* the *calorific value* and the *heating value.* The quality of a material to be used as a fuel is determined by its heat of combustion. The greater the heat of combustion, the more intense is the heat released during the combustion, and the higher the temperature of the flames. The values of the heat of combustion are intensively used for the calculation of the fire load, that is the maximum heat that would be released in a fire if all the combustibles are completely burned (see subsection 2.1.4).

Heats of combustion are tabulated in corresponding handbooks. For the common solid materials they are in the range 2–50 MJ/kg. Exceptions are some materials which are known as very efficient fuels. For example, for diborane, H_6B_2, and hydrogen, H_2, these values are 79.80 and 141.79 MJ/kg, respectively.

Heats of combustion are determined in laboratories by various calorimetric techniques.

While heats of combustion vary considerably for different materials, a derived value, that is the heat of combustion per kg of oxygen consumed, is nearly constant for the majority of organic materials. Huggett (1980) has shown that for the stoichiometric fuel-to-oxygen mass ratio this value is equal to 13.1 MJ/kg O_2. The supposition of the constancy of the heat of combustion per kg of oxygen is employed in several laboratory techniques for determining the heat release rate (see *Rate of heat release*, this subsection).

Oxygen index

A useful method for the quality control of production processes and products has been the determination of the oxygen index of a material. The basis of this method is in finding the minimal concentration of oxygen in a nitrogen/oxygen mixture which will just support unaided burning of a candle-like specimen of the material. Since the normal percentage of oxygen in air is 21%, it follows that all the materials which freely burn in the air have an oxygen index of at least 21, or, more likely, below this value. Obviously, the lower the oxygen index, the more flammable is the material, and vice versa.

There is a pronounced bias of experimenters to predict the fire behavior of a material on the basis of its oxygen index value. However, fire practice shows that while this may be true for some materials, gross deviations are quite frequent. For example, some species of wood, leather, and polyvinyl chloride all have oxygen index values higher than 30, meaning that for their free combustion an atmosphere highly enriched in oxygen would be needed, which is, of course, not the case in real fires. Standards for oxygen index determinations invariably stress that test results cannot be used for the prediction of the material's behavior in actual fires.

Other fire characteristics of solids

There are a number of other characteristics which are useful for the prediction of the fire behavior of a solid material (GOST 12.1.017-80 1980). These will be briefly commented on, as a detailed discussion is outside the scope of this book.

Ignition temperature. The ignition temperature of a material is defined as the minimum temperature at which flammable vapors and gases are released from the material at such a rate that a self-sustained combustion is initiated after the removal of the external pilot source. The principal variables that influence values of ignition temperature are the rate of oxidant supply (air flow), the rate of heating and the physical state and geometry of the solid.

Values of this parameter are useful in the prediction of fire hazards in the process industries.

Autoignition temperature. The autoignition temperature of a material is the lowest temperature at which there is a sudden increase in the rate of the exothermic reaction which ends up in a flaming mode of combustion. As the definition states, no pilot ignition source takes part in the process of flammable combustion.

The autoignition temperature is not a reliable indicator of the fire hazard of a material since it is influenced by many factors, such as the environmental conditions of temperature and pressure, the rate and duration of heating, compatibility or incompatibility with other materials with which the solid may be mixed, the possibility of the catalytic effect of other materials, etc.

Self-heating temperature. The self-heating temperature of a material is defined as the temperature at which practically different exothermic processes begin to appear. In industry it is used for the selection of nonhazardous conditions of heating of a material. It is unwise and hazardous to heat continuously a material at temperatures close to its self-heating temperature. Usually, a maximum temperature, equal to 90% of the self-heating temperature, is set and kept to.

Melting temperature. The melting temperature of a material is defined as the temperature at which there is a sudden increase in the exothermic reaction which ends up in the appearance of the melt. This parameter is useful for the fire characterization of thermoplastics, for the fire investigation and expertise and for fire prevention in the process industries. Usually, the melting temperature of a solid is very close to or equal to its ignition temperature.

Effects of water, oxygen and other materials. When different kinds of materials are stored together, either in the open or in warehouses, it is important to have some information on the possible effects of water, air or other materials on the stored goods. The same information is required when designing fire- and/or explosion-safe industrial processes.

The methodology of the determination of this parameter is rather simple. On a given amount of the material, water is added, or the material is intimately mixed with another material and the temperature of the mixture recorded. If the temperature of the mixture after some time does not rise more than 0.5 K above room temperature, it may be concluded that the solid under investigation can be safely stored or processed with another material, or that contact with water or air will not be dangerous. If not, then experiments for determining the self-ignition temperature should be conducted.

Minimal concentration of extinguishing agent. The minimal concentration of an extinguishing agent in a mixture with air refers to that concentration which makes possible a momentary extinguishment of the diffusion flame of the material on fire. It is used in the design of fixed fire extinguishing systems.

Compatibility with water-based foams. Some solids when heated to higher temperatures easily melt before catching fire. Since water-based foams are the choice agent for extinguishing fires of liquids it is important to know whether the addition of foam on the liquefied material is safe or not. The effect of the contact of the foam with the liquid is visually followed. If on the contact of the foam with the material there is an explosion, or a vigorous reaction followed by splashing and frothing of the liquid form of the material, such a foam should obviously not be used for extinguishing such fires.

Sensitivity on impact. Information on the sensitivity of a material on impact is used for the safe production, application, storage and transport of materials.

2.1.1.2 Fire hazards of the most common solids

The hazards of the most common solid materials will be evaluated on the basis of the above fire characteristics.

Wood and wood-based products

Wood and numerous wood-based products, such as paper, cellulose-based fibrous materials, and many others, are indeed ubiquitous. They are invariably involved in almost all kinds of fires. Therefore, understanding of their fire characteristics is important for fire protection.

The chemical content of dry wood and wood-based products is relatively simple. Carbon (*ca.* 50%), oxygen (*ca.* 40%), and hydrogen (*ca.* 6%) are the most abundant elements. Nitrogen and mineral ash make up the remainder. However, these few elements are combined to form a large number of substances, of which cellulose (50%), lignin (26%) and extractables (1%) are the major components. Wood also contains water, either as free water (moisture) or as absorbed water in wood cells. Whereas moisture is readily removed on heating wood and wood products above 380 K, absorbed water remains even after prolonged heating. Apparently, dry wood may still contain considerable amounts of water (5–6%).

Wood and wood-bassed products are combustible. They can burn in different modes, such as smoldering, charring, ignition followed by flames, and burning with a lot of smoke. This suggests that a number of variables profoundly influence the burning behavior of wood. Among them the most significant are:

— the physical form. Wood dust is capable of explosive detonation, small pieces of wood in the form of kindling readily ignite, whereas wood-logs burn with difficulty and only after an extended exposure to heat. These differences are explained by the well known effect of increasing the surface area-to-volume ratio as the size of the particles is diminished. The larger the surface area, the greater is the exposure of wood to the air and the lesser the mass to conduct heat away from the particles.

— the product $K\rho C$ (thermal absorptivity, see section 4.2) for wood, compared with

other solids, is rather low. This makes wood a good insulator and contributes to relatively ready ignition.
— the moisture content: the greater the moisture content, the more difficult it is to ignite wood. The moisture content of wood is in a reversible equilibrium with the moisture in the air (relative humidity). Big conflagrations, in which large areas under grass, wood or bush are set on fire, usually occur following spells of dry weather.

Consequently, the ignition temperature of wood is variable. Normally, values between 460 and 660 K are measured.

Wood and wood-based products generally do not spread flame quickly. Rates between 0.1 and 0.3 m/s have been measured, with results again dependent on a number of factors.

The heat of combustion for wood is typically 18–21 MJ/kg. Thus, wood is not a very calorific fuel.

The rate of heat release of wood depends primarily on the intensity of the incident radiant heat flux. Values of 90–170 kW/m^2 have been measured (Parker & Long 1972).

Smoke produced from burning wood has a characteristic recognizable odor. In comparison with other solids the toxicity of smoke produced from wood burning is not pronounced. Except for carbon monoxide, other toxic gases are either absent or present only in traces.

Plastics

Plastics are used practically everywhere: in building construction, homes, offices, shops, schools, hospitals, etc. A lot of equipment and applications, furniture, wall coverings, curtains, textiles and many other products either are made completely of plastic, or contain some plastic parts.

All plastics, regardless of chemical characteristics, being organic compounds are combustible. Various flame retardants can considerably reduce flammability, but cannot completely stop combustion. Ignition temperatures of plastics are somewhat higher in comparison with wood and wood-based products. However, the rates of flame spread are generally much higher, and under favorable conditions may reach 0.5–0.6 m/s.

The burning of plastics rapidly produces smoke which is usually dense, contains a lot of soot and has a dark color. It has been found in many cases that the inhibition of flammability of plastics by flame retardants increases smoke production.

The oxygen index of plastics ranges from 16.1 for polyurethane foam, 17.4 for polyethylene, 18.7 for polyrayon, 20.6 for polyester fibers, 22.6 for polyvinyl fluoride, 26.8 for modacrylic fibers, 29.0 for polyphenylene oxide to 37.4–49 for PVC, and 95 for polytetrafluoroethylene (Babrauskas 1986b).

Thermoplastics soften on heating before reaching ignition temperature and harden on cooling. The process of heating and cooling can be repeated. At higher temperatures, thermoplastics melt and flow. This characteristic of plastics is potentially hazardous since the flaming liquid may drip and thus spread fire.

During the burning, some plastics release corrosive and toxic gases such as HCl, HF, HBr, HCN, and NH_3. The conditions which enhance the emission of such gases during fire are low ventilation and lengthy fire growth which increase fire temperatures to the point of an easy break of the polymer matrix and the release of simple gaseous constituents. For example, the noticeable emission of HCl from PVC starts at 523 K and is very fast at 723 K (Smith 1985). The release of HCl from PVC treated with $CaCO_3$ and other basic substances as a filler is considerably smaller (Smith 1985, Brown & Martin 1986). The degradation of plastics containing fluorine begins at 573 K and is fast at temperatures above 773 K (Smith 1985). Gaseous NH_3 evolves from nylon between 658 and 678 K (Humphreys & Christopher 1985). Only 3–33% of the nitrogen in a nylon molecule evolves as gaseous NH_3. The remaining nitrogen is released in the form of other nitrogenous products, while almost 50% of the nitrogen is retained in the charred residue (Humphreys & Christopher 1985).

The release of HCN from flexible polyurethane is strongly dependent on the burning mode. A two-step combustion of polyurethane, that is smoldering followed by flaming, is especially hazardous, since the concentration of evolved HCN is 10 times greater than the concentration of the same gas released during a one-step combustion mode, either smoldering, or flaming (Levin *et al.* 1985). Fire-retardant-treated polyurethane emits more HCN than an untreated one. This finding points to the need for specifying the combustion mode of samples when determining and reporting the toxicity of plastics.

A 0.018–0.034 $kg/m^2/s$ combustion rate for plastics has been measured (Babrauskas 1986c).

The behavior of rubber in fire is similar to that of some plastics, except that the rate of burning is more moderate. The combustion of rubber proceeds with the evolution of dense smoke with a characteristic smell. Rubber in the form of foam may ignite spontaneously, especially when freshly stored and used.

It must again be stressed that combustibility characteristics and suitability for use should be determined for any material or assembly on the basis of tests which simulate the end-use conditions as realistically as possible.

Textiles

The widespread use of textiles in daily life, coupled with the fact that nearly all textiles are combustible, explain the leading role of textile fires in fire deaths. More than 50% of fatal incidents involve a fabric (Krasny & Sello 1986). As to the type of fabric first ignited, artificial fibres, cotton and rayon comprise the largest percentage (41%), whereas wool and wool mixtures are very rarely first ignited (only *ca.* 1%). This marked difference is due to the differing ignition temperatures. While cotton and most artificial fibres ignite at relatively low temperatures (520–670 K), the ignition temperature for natural protein-based fabrics, such as wool, silk and cashmere, is between 840 and 880 K.

The fire chracteristics of textiles depend on the nature and proportion of individual fibers, on their weight and the method of blending. Textiles based on cellulosic fibers (cotton, jute) behave differently in fires compared to protein fibres, e.g. hair fibres derived from some animals. The latter ones do not burn readily, shrink at temperatures approaching their decomposition temperature, and burn more slowly. Common artificial fibers (acetate, nylon, polyester, rayon, etc.) burn similarly to protein-based natural fibers. They are thus relatively safer compared to fabrics containing cellulose. However, when exposed to heat, artificial fibers, being thermoplastic, often melt and stick to the skin. Therefore, they should not be employed for protective clothing.

The oxygen index of textiles is often low. This means that after the ignition source is removed, textiles will continue to burn even in atmospheres poor in oxygen. Ideally, textiles should have a high oxygen index, have a high ignition temperature, should release little smoke and toxic gases, be self-extinguishing or spread slame slowly, generate little heat and should not melt.

The fire characteristics of textiles should be tailored to their end use, that is whether for clothing, wall coverings, curtains, carpets, etc. In many countries textiles intended for clothing must past rather severe fire tests. These tests rule out all extremely flammable fabrics.

One proven way of improving the fire characteristics of flammable textiles is to use flame retardants. Textiles treated with flame retardants are more difficult to ignite, and if they do ignite, they burn more slowly. Fire and flame retardant chemicals are normally added to textiles during the manufacturing process. The effectiveness of a fire retardant is commonly checked by the oxygen index test.

Synthetic fabrics can be hazardous in some special atmospheres, such as oxygen-enriched atmospheres, or atmospheres containing flammable gases and vapors, because of the accumulation of static electricity. The discharge of this electricity to ground or other objects may produce a spark of sufficient energy to ignite a flammable gas. In such situations, electrically conductive fabrics should be employed.

Chemicals

Fires involving chemicals, especially gases, can be disastrous. Marshall (1987) has thoroughly analyzed some great fires caused by liquefied natural gas (Cleveland, Ohio, 1944), liquid propane (Feyzin, France, 1966), cyclohexane (the Flixborough, England, 1974, accident), liquid propylene (San Carlos, Spain, 1978) and liquefied petroleum gas (Mexico City, 1984).

Chemical hazards are in many ways unique. Chemicals which are produced, stored, shipped and used commercially differ from other materials in that they frequently pose several hazards simultaneously. In addition to fire and explosion hazards, many chemicals are toxic, unstable, react vigorously with air and/or water, and are able to corrode and oxidize other materials. This means that even if a chemical is not combustible *per se*, it can still be hazardous in a fire environment. Each of these hazards will be briefly analyzed. Although, strictly, only solid chemicals should be discussed in this subsection, chemicals appearing in the other two aggregate states (liquid and gaseous) will also be mentioned to avoid later repetition of their general characteristics.

Toxic chemicals. The toxic hazards of a chemical result from inhalation, skin adsorption or oral ingestion during spills, leaks or fires. Information on the toxicity of chemicals can be found in the better chemical dictionaries, encyclopedias and handbooks. The severity of a toxic hazard has been designated (NFPA 49 1975) by five numerical grades, from (0) — no unusual hazard, (1) — slightly hazardous to health, (2) — hazardous to health, (3) — very hazardous and (4) — extremely hazardous. With the last category, normal protective clothing does not provide adequate protection. The extremely toxic chemicals are shown in Table 2.1.

Table 2.1 — Some extremely toxic chemicals (belonging to grade (4) in the NFPA 49 grading system)

Chemical	Formula	Aggregate state
Inorganic		
Beryllium	Be	Solid powder
Bromine	Br_2	Reddish-brown liquid or vapor
Bromine pentafluoride	BrF_5	Colorless fuming liquid
Bromine trifluoride	BrF_3	Colorless fuming liquid
Chlorine trifluoride	ClF_3	Greenish-yellow liquid
Cyanogen	NCCN	Colorless gas
Fluorine	F_2	Pale-yellow gas
Hydrogen cyanide	HCN	Colorless liquid or gas
Hydrogen fluoride	HF	Colorless liquid, solution or gas
Nickel carbonyl	$Ni(CO)_4$	Colorless liquid
Phosgene	$COCl_2$	Colorless gas
Organic		
Adiponitrile	$NC(CH_2)_4CN$	Colorless liquid
Benzotrifluoride	$C_6H_5CF_3$	Colorless liquid
3-Bromopropyne	$CH{\equiv}CCH_2Br$	Colorless liquid
Chloropicrin	CCl_3NO_2	Colorless liquid
Dimethyl sulfate	$CH_3OSO_2OCH_3$	Colorless liquid
2-Hydroxy-2-methyl-propanenitrile	$(CH_3)_2C(OH)CN$	Colorless liquid
Methyl parathion	$(CH_3)_2P(S)OC_6H_4NO_2$	White solid
Parathion	$(C_2H_5O)_2PSOC_6H_4NO_2$	Brownish liquid

Oxidizing chemicals. The primary reason for the hazard of oxidizing chemicals stems from their richness in oxygen. The 'excess' of oxygen comes in the form of groups such as peroxide, O_2^{2-}, nitrate, NO_3^-, nitrite, NO_2^-, perchlorate, ClO_4^-, chlorate, ClO_3^-, chlorite, ClO_2^-, hypochlorite, ClO^-, dichromate, $Cr_2O_7^{2-}$, permanganate, MnO_4^- and persulfate, $S_2O_8^{2-}$. At higher temperatures, and for some materials even at normal temperatures, these chemicals liberate oxygen which strongly increases the fire intensity.

Mixtures of oxidizing chemicals and ordinary combustible materials, such as wood, textiles, paper, plastics, etc., are hazardous even when no flame ignition source is present. These mixtures can be ignited by heat energy originating from a very weak ignition source (friction, impact or static electricity). Since most oxidizing chemicals are of inorganic origin they are not combustible in themselves, which, however, does not significantly reduce the general hazard of oxidizing chemicals.

Some common oxidizing chemicals (NFPA 49 1975) are shown in Table 2.2. Also

Table 2.2 — Common oxidizing chemicals

Chemical	Formula	Aggregate state
Inorganic		
Nitrates	$M(NO_3)_n$, $n=1–4$ $M=NH_4^+$, Ba^{2+}, Co_2^+, Cu^{2+}, Pb^{2+}, Mg^{2+}, Ni^{2+}, K^+, Ag^+, Na^+, Sr^{2+}, Th^{4+}, UO_2^{2+}, H^+	Solids, except HNO^3 (liq.)
Perchlorates	$M(ClO_4)_n$, $n=1,2$ $M=NH_4^+$, Mg^{2+}, H^+, Na^+	Solids, except $HClO_4$ (liq.)
Chlorates	$M(ClO_3)_n$, $n=1,2$ $M=Ca^{2+}$, K^+, Na^+, Zn^{2+}	Solids
Chlorites	$NaClO_2$	Solids
Hypochlorites	$Ca(ClO)_2$	Solids
Bromates	$KBrO_3$	Solids
Peroxides	$M(O_2)_n$, $n=1,2$ $M=Ba^{2+}$, H^+, K^+, Sr^{2+}	Solids, except H_2O_2(1)
Dichromates	$Cr_2O_7^{2-}$	Colored solids
Permanganate	$KMnO_4$, NH_4MnO_4	Dark solids
Persulfate	$K_2S_2O_8$	White solid
Bromine	Br_2	Liquid or vap.
Bromine pentafluoride and trichloride	BrF_5, $BrCl_3$	Colorless liq.
Chlorine	Cl_2	Gas
Chlorine trifluoride	ClF_3	Greenish liq.
Chlorosulfuric acid	ClSOOH	Colorless liq.
Chromic trioxide	CrO_3	Dark-red solid
Fluorine	F_2	Yellowish gas
Nitrogen peroxide	$N_2O_4(NO_2)$	Brownish gas
Nitrogen trioxide	N_2O_3	Unstable liq.
Liquid oxygen	O_2	Blue liquid
Organic		
Amyl nitrate	$CH_3(CH_2)_4NO_3$	Colorless liq.
t-Butyl hydroperoxide	$(CH_3)_3COOH$	Colorless liq.
t-Butyl perbenzoate	$C_6H_5COOOC(CH_3)_3$	Colorless liq.
t-Butyl peroxyacetate	$CH_3CO(O_2)C(CH_3)_3$	Liquid
t-Butyl peroxypivalate	$(CH_3)COOCOC(CH_3)_3$	Colorless liq.
Dibenzoyl peroxide	$(C_6H_5CO)_2O_2$	White solid
t-Dibutyl peroxide	$(CH_3)_3COOC(CH_3)_3$	Colorless liq.
Diisopropyl peroxydicarbonate	$(CH_3)_2CHOCO–OOCO–OCH(CH_3)_2$	Solid
Dilauroyl peroxide	$(C_{11}H_{23}CO)_2O_2$	White solid
n-Propyl nitrate	$CH_3(CH_2)_2NO_3$	Liquid
Chloro isocyanurates	$Cl(NCO)_3^{2-}$	White solids

included in this table are chemicals which, although containing no oxygen atoms, are oxidants according to the broader definition of oxidants (that is, compounds that readily take up electrons).

Combustible chemicals. NFPA 49 (NFPA 49 1975) rates chemicals as either noncombustible, grade (0) in the diagram for hazard identification, or combustible with different intensitics (gradcs (1) to (4)).

Most solid chemicals are rated (1), that is they burn only after a prolonged exposure to fire. However, solids in the form of small particles and pieces, for example coarse dust, scrap, and solids which tend to self-ignite, are more hazardous (rated as grade (3)). Even more hazardous are some solids in the form of fine dust which are categorized in grade (4).

Liquid chemicals are rated as either grade (2) (must be moderately heated before ignition), grade (3) (ignite under normal conditions of temperature and pressure), or grade (4) (vapors are readily flammable and explosive).

The grade (4) chemicals are shown in Table 2.3.

Table 2.3 — The most readily flammable chemicals (grade (4) in the NFPA 49 grading system)

Solids	Ethylchloride
Calcium carbide	Ethyl methyl ether
Lithium hydride	Ethyl nitrite
Powdered nickel	Isoprene
Zirconium	Isopropylamine
	Methylformate
Dibenzoyl peroxide	Propylene oxide
Picric acid	Trichlorosilane
Trinitrobenzene	
Trinitrotoluene	**Gases**
	Acethylene
	1,3-Butadiene
Liquids	Butane
Liquid hydrogen	Butenes
Hydrogen cyanide	Cyanogen
Liquefied natural gas	Cyclopropane
	Dimethyl ether
Acetaldehyde	Ethene
t-Butyl hydroperoxide	Ethylene oxide
Butyllithium	Formaldehyde
1,1-Dichloroethene	Methylamines (mono-, di-, tri-)
Diethyl ether	Methyl chloride
Diisopropyl peroxidicarbonate	Propylene
Dimethyl sulfide	Tetralfuoroethylene monomer
Ethylamine	Vinyl chloride

Reactive chemicals. Almost all chemicals are hazardous enough to warrant precaution in handling, transporting or storing them. The reactivity of chemicals ranges from 'normally stable' chemicals (NFPA 49 grade (0)) to very reactive chemicals which radily detonate or decompose at normal temperatures and pressure (grade

(4)) in the presence of a weak ignition source. Many of the most reactive chemicals are frequently also very flammable and toxic, or are strong oxidizing agents.

The most reactive chemicals are shown in Table 2.4.

Table 2.4 — The most reactive chemicals (NFPA 49 grade (4))

Solids	**Liquids**
Ammonium perchlorate	3-Bromopropyne
	t-Butytl perbenzoate
	t-Butylhydroperoxide
1-Chloro-2,4-dinitrobenzene	*t*-Butyl peroxyacetate
Dibenzoyl peroxide	Cumene hydroperoxide
Diisopropyl peroxyhydrocarbonate	Diacetylperoxide
o-Dinitrobenzene	*t*-Dibutyl peroxide
Picric acid	Ethyl methyl ketone peroxide
Trinitrobenzene	Ethyl nitrite
Trinitrotoluene	Peroxyacetic acid

Air and water reactive chemicals. Some chemicals are unstable in air or in contact with water. Hazards of such chemicals result from any one or more of the following:

— liberation of heat leading to local overheating which may ignite the chemical itself (if flammable), or nearby combustible matertial (for example, a flammable solvent),
— release of flammable gases, such as hydrogen or acetylene,
— liberation of strong oxidizing gases (Br_2, F_2, Cl_2, NO_2),
— release of fumes of metal oxides (for example, ZnO),
— liberation of toxic gases (H_2S, SO_2),
— liberation of corrosive acids (HF, HCl) and bases (NH_3).

Some water reactive chemicals, together with principal hazardous product(s) of reaction, are shown in Table 2.5 (NFPA 49 1975).

Corrosive chemicals. Many chemicals listed as oxidizing or water and air reactive are at the same time corrosive. That is, they may damage living organisms upon contact, inhalation or ingestion. Most inorganic acids and all of the halogen elements are corrosive.

Metals

Fire hazards of metals arise almost exclusively in the industrial environment. Compared with other types of materials, fires of metals occur less frequently.

Although it is true that the majority of metals are combustible, metals can be set on fire only under some very specific conditions. Thus, the phsyical form of a metal, and the size and shape of its particles, strongly determine the ease of ignition. Metals

Table 2.5 — Chemicals reactive with water

Solids	Product of reaction
Aluminum chloride (anhydr.)	Hydrogen chloride
Calcium	Calcium hydroxide, hydrogen
Calcium carbide	Acetylene
Lithium	Lithium hydroxide, hydrogen
Lithium hydride	Lithium hydroxide, hydrogen
Lithium tetrahydroaluminate	Lithium hydroxide, hydrogen
Magnesium and Mg alloys	Hydrogen
Phosphorus trichloride	Phosphorus acids, hydrochloric acid
Phosphorus pentasulfide	Hydrogen sulfide
Phosphoryl chloride	Hydrogen chloride
Potassium	Heat, hydrogen, potassium hydroxide
Potassium peroxide	Heat, hydrogen, potassium hydroxide
Sodium, Na	Heat, hydrogen, sodium hydroxide
Sodium peroxide	Heat, hydrogen, sodium peroxide
Sodium–potassium alloys	Heat, hydrogen, sodium peroxide
Chlorine trifluoride	Hydrogen fluoride, chlorine
Chlorosulfuric acid	Hydrogen chloride, sulfuric acid, hydrogen in contact with metals
Fluorine	Hydrogen fluoride
Sulfuryl chloride	Sulfur dioxide, chlorine
Thionyl chloride	Sulfur dioxide, chlorine, sulfur monochloride, hydrogen chloride
Trichlorosilane	Hydrochloric acid
Vanadium tetrachloride	Hydrochloric acid

are most hazardous when in their final divided form, such as dust clouds. In fact, most metal dusts when ignited lead to explosions. Other hazardous forms of metals are fine, freshly made particles, and metals in molten form.

Principal hazards that result from metal fires, in addition to the usual hazards of toxicity of combustion products, include very high flame temperatures, the possibility of steam and hydrogen explosion and explosion in contact with some extinguishing agents.

The extinguishment of some metal fires remains something of a problem even for experienced firefighters. Agents which are successful in extinguishing fires of ordinary combustibles and liquids may be useless with metals, because some metals continue to burn in an inert atmosphere of carbon dioxide or nitrogen.

With respect to fire characteristics, metals can be divided into two broad classes: combustible metals, such as sodium, Na, potassium, K, magnesium, Mg, calcium, Ca, zinc, Zn, hafnium, Hf, plutonium, Pu, thorium, Th, titanium, Ti, uranium, U, and zirconium, Zr, which ignite relatively easily, and metals which can burn only with difficulty, notably aluminium, Al, and iron, Fe. The ignition temperatures of solid metals are rather high. Heats of combustion rarely exceed 20 MJ/kg. Only beryllium, Be (66.4 MJ/kg), aluminum (31.0 MJ/kg) and magnesium (24.7 MJ/kg) can be considered energetically rich, while the heats of combustion for all other metals are lower, between 2.0 and 20.0 MJ/kg (Babrauskas 1986b).

Another hazard of metals in a fire environment comes from the marked inability of buildings and other constructions made of metals, such as iron (steel) and aluminum, to withstand temperatures within the body of construction higher than a few hundred K. Such relatively low temperatures are easily reached in most fires after just a few minutes. On further temperature increase, unprotected metal structural elements lose their load bearing capacity which may cause collapse of the structure.

2.1.2 Liquids

Flammable and combustible liquids are among the most fire hazardous materials. Fire statistics record that fires involving, or started with the involvement of, a liquid are the most frequent ones. In industry, over 20% of all fires are caused by a liquid. Flammable and combustible liquids are the most hazardous in all instances when the liquid is exposed to air (spillage, accidental overturning of containers, leakage during transportation or handling).

The fire and explosion characteristics of a liquid can be described using a number of parameters. Some of them are applicable to solid and gaseous materials as well. Thus, the flammability group, ignition temperature, self-ignition temperature, hazard of explosion and/or fire in contact with water, oxygen in air and other materials, and the minimal concentration of a fire extinguishing agent described in section 2.1 may also be employed for the fire characterization of liquids. In addition there are some characteristics which can be used specifically for fire charaterization of liquids. Among them, the most important are as follows:

Combustibility potential

The combustibility potential of a liquid is a quantitative measure of the combustibility of a liquid. It is defined as the difference in energy required for a sustained burning of a liquid and the energy which is released in burning the optimal composition of vapor–air mixture of the liquid. The combustibility potential is determined from experimentally obtained flammability diagrams. For liquids that are flammable or combustible in a given environment it has a negative value, while for noncombustible liquids its value is positive.

The combustibility potential is used for the calculation of both the lower flammability limits of liquids and the minimal concentration of an extinguishing agent needed for the extinguishment of the fire of the given liquid.

Flash point

The flash point is a very useful fire characteristic of a liquid. It may be defined as the lowest temperature of the liquid at which a mixture of vapor and air is formed which, when ignited, will just support the ignition, with no flame spreading away from the ignition source. If corresponds approximately to the lower flammability limit of the given liquid.

The flash point of a liquid is not a constant. It varies owing to many factors such as air pressure, oxygen content in air, the purity of the liquid being tested and the method of determination. There are two principal test methods for flash point determination, related to the kind of test apparatus employed: the closed-cup and the open-cut methods. The closed-cup flash points are determined using the Tagliabe Closed Tester and the Pensky–Martens Closed Tester, while the open-cup flash points are determined using the Taglıable Open Cup and the Cleveland Open Cup testers. Flash points determined by the closed-cup methods are generally lower than the open-cup flash points since flammable mixtures are formed more easily in closed vessels than in the open.

The flash point is commonly used, together with the vapor pressure and the boiling point, as the most important criterion for fire classification of liquids. Liquides that have flash points below 311 K and boiling points at or above 311 K are classified as flammable, while liquids with flash points at or above 311 K are designated as combustible. Flammable liquids correspond to Class I, while combustible liquids comprise Class II and Class III liquids. Classes I and III are further divided into classes A, B or C. A graphical presentation of the classification of flammable and combustible liquids is shown in Fig. 2.2.

Lower and upper flammable limits

A liquid will burn only when its vapor, mixed with air in certain proportions, is ignited. The limits of the proportions, usually expressed in percentages by the volume of vapor — or gas — in air, are termed the 'lower' and 'upper' flammable limits. Thus, the lower flammable limit is the minimum concentration of vapor in air below which it is not possible to ignite the mixture in a manner that flames can be propagated. Likewise, the upper flammable limit gives the maximum concentration of vapor in air above which flame propagation cannot happen. In the first case the mixture is said to be too 'lean' to burn, while in the second case the mixture is described as too 'rich' to burn. For the majority of common liquids, the lower flammable limit is around 2–3%, while the upper flammable limit is around 10%.

Information on flammable limits of a liquid is important for the design of fire-. and explosion-safe conditions in the process industries, in ducts, in ventilation systems, and in working environments where potential ignition sources may be present. In this respect it is even more important to have knowledge of the effects of temperature and pressure on flammable limits, since most tabulated data on flammable limits refer to normal temperature and atmospheric pressure. It has been found that the limits may be extended if the temperature or pressure is increased, and contracted if the temperature or pressure is decreased.

Minimal energy of ignition

The minimal energy of ignition of the vapors of a liquid is defined as the lowest energy of the electrostatic discharge capable of igniting the optimal mixture of vapor with air. As the composition of a flammable mixture approaches either end of the flammable range, the ignition energy becomes greater, that is the mixture becomes more and more difficult to ignite by electrostatic discharge.

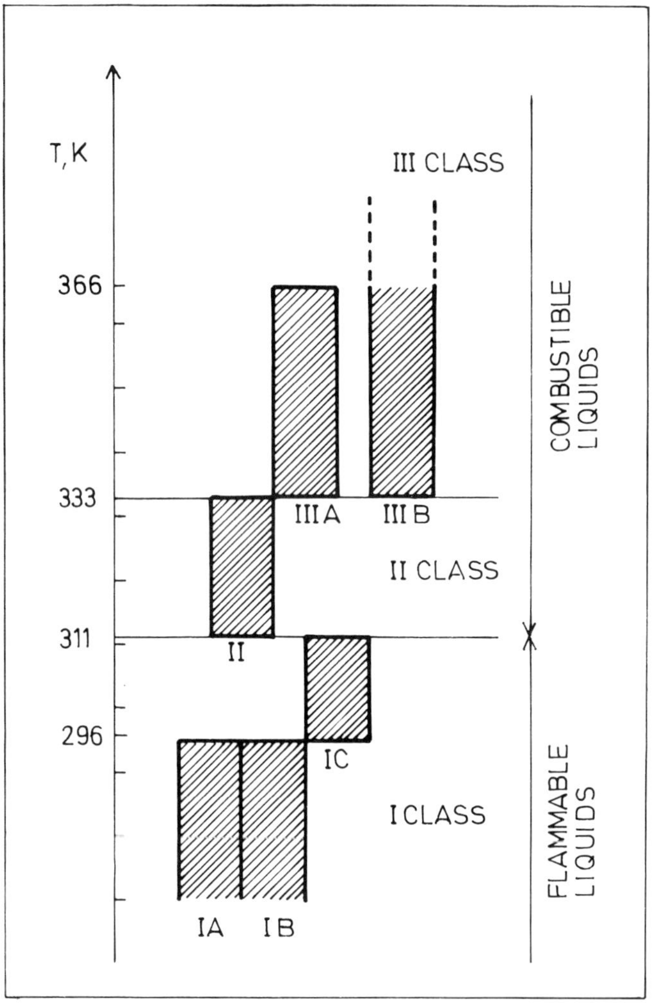

Fig. 2.2 — Classification of liquids on the basis of flash points.

Burning rate

The burning rate of a liquid may be defined as the amount of liquid which is burnt in a unit of time from a unit of area. This fire characteristic of liquids is usually employed for calculating the duration of combustion in liquid reservoirs, intensity of heat release and temperature regimen of a fire.

For pure, compositionally homogeneous liquids, the burning rate is constant. For mixtures of liquids the burning rate is changeable, as lighter fractions burn more rapidly than heavier fractions. Typical burning rates are 10 to 30 cm of liquid depth per hour. Of course, a number of other factors, such as environmental factors and latent heat of varporization, can influence the burning rate.

In addition to the above pure fire characteristics, many physical properties of liquids are important for understanding the fire behavior of liquids. For example, the specific mass of a liquid can point to possible ways of extinguishing liquid fires with water and water-based foams. Liquids which are insoluble in water and are heavier than water will sink below the water surface on contact with water, whereas liquids with a specific gravity lower than that of water ($\rho=1$) will float on the surface.

Another physical property of interest in the evaluation of fire hazards of a liquid is the vapor density ratio. It is defined as the ratio of the relative mass of a volume of vapor to the mass of an equal volume of air under the same environmental conditions. Liquids with vapor densities greater than that for air ($\rho=1$) will tend to settle and concentrate on the bottom.

Other physical properties of a liquid, such as the vapor pressure, evaporation rate, latent heat of vaporization, solubility in water and viscosity, when properly evaluated may be of use in fire protection. Liquids which possess greater vapor pressure, have greater evaporation rates, have lower heats of vaporization and are insoluble in water are inherently more fire hazardous.

2.1.3 Gases

For fire protection purposes a gas may be defined as any substance which exists in a gaseous state at normal temperature and pressure. Since at these conditions many substances may exist either as liquids or as gases, depending on the partial pressure of their vapors, it is generally accepted that all those liquids which exert a relatively high vapor pressure, exceeding 275 kPa at 311 K, may be regarded as gases.

In most situations, gases are used in large volumes. Since gases are much lighter than liquids, and lighter still than solids, the only practical means of having a reasonable quantity of a gas on hand is either by gas compression in containers or by filling the containers with liquefied gas. Both forms of gas packing present hazards.

Fire and explosion hazards of gases are very similar to those of liquids. This is not surprising since in fire hazards of liquids it is the vapor of a particular liquid which is hazardous, rather than the liquid phase itself. This also means that most of the fire and explosion characteristics described in subsection 2.1.2, which refer to the hazards of liquids, are equally applicable to gases. The exception is the term 'flash point'. Although this is a useful characteristic in the case of liquids, its definition is of no use for flammable gases, since flammable mixtures of gases and air are formed in all cases at temperatures well above the normal boiling point of most gases.

In addition to the fire characteristics, many gases are hazardous on account of other properties. The hazard stems from properties such as toxicity, reactivity and chemical inertness. However, the basic hazard of all gases and vapors, regardless of their chemical composition, is related to the change of gas pressure with changing temperature. The classical gas law, $P_0V_0/T_0=P_1V_1/T_1$, when applied to gases confined in a container ($V_0=V_1$) reduces to $P_0/T_0=P_1/T_1$ (subscript 0 refers to initial conditions, and subscript 1 refers to new, changed, conditions). Thus, temperature doubling leads to a doubling in gas pressure. This means that gas containers would normally be ruptured in real fires were they not supplied with overpressure relief devices. If the compressed or liquefied gas is highly toxic or reactive, overpressure relief devices are not installed on containers since it is felt that the hazard of gas release indoors, following operation of the relief device, outweighs the hazard of container failure due to overpressure.

A special type of hazard exists when containers of liquefied gases rupture for some reason, such as the engulfment in fire, corrosion or impact from a crash. The resulting pressure release is manifested in the form of explosions which are often termed the BLEVE (Boiling Liquid Expanding Vapor Explosion). The cause of a BLEVE may be inferred from its acronym. Basically, when a partially filled container containing a liquefied gas under pressure fails for mechanical reasons the pressure within the container is reduced to atmospheric. Since containers are normally stored or handled at ambiental temperatures which are well above the boiling point of the liquefied gas there is a spontaneous tendency for the liquid phase to boil off the upper portions of the liquid phase and thus form large quantities of the vapor phase. The temperature differences are normally such that about one third of the liquid in the container becomes vaporized. The remaining liquid phase is cooled down to temperatures close to its normal boiling point. The vapor formed in the process of sudden liquid evaporization exerts a strong pressure which may propel ruptured pieces, and manifests all other effects of an explosion.

The situation gets worse when the container is engulfed in fire. In this case the temperature difference between the liquid at the instant of container failure and the normal boiling point of the liquefied gas is even greater. Also greater are the portions of the liquid which are vaporized (often more than 50% of the starting quantity) and the quantities of vapor thus formed. When the vapor is mixed with air within the flammable limits and is ignited by fire flames the mixture explodes into the characteristic fireball which may reach greater proportions. Although one might think that a functioning relief valve would prevent the container exceeding its designed pressure, the partially filled container is bound to fail owing to different heat transfer regimes of the container walls in contact with gas or vapor and with the liquid gas (Marshall 1987). Portions of the walls at, and below, the liquid level cannot reach temperatures higher than the boiling temperature of the liquid owing to the high heat capacity of the liquid phase. However, the walls above the liquid phase are not kept at constant temperature since there is no longer a boiling liquid which would conduct away the heat, and the wall temperature rapidly reaches flame temperature. At these temperatures, iron and steel are so weakened that the container is easily ruptured, causing the BLEVE.

The consequences of a BLEVE can be very destructive, as examples of some accidents reported in the literature demonstrate. Marshall (1987) thoroughly analyzed the 1966, Feyzin, France, BLEVE, an accident in which five spherical tanks containing liquid propane exploded and caused a conflagratory fire involving a number of petrol and crude oil tanks. The BLEVE killed 17 people and injured many more. Marshall criticized the term BLEVE as being imprecise and of uncertain meaning, and recommended that it no longer be used.

2.1.4 Dusts

It has already been shown (subsection 2.1.1) that many fire characteristics of flammable and combustible solids are strongly dependent on the particle size of the particular material. Generally, the smaller the particle size, the greater are the fire and explosion hazards of that material. The effect is easily explained by the fact that the surface-to-volume ratio increases with decreasing particle size. Dusts therefore have a large surface area, which means that the reaction rate in fires is increased, often to the point of a deflagration. Many materials in bulk or in the form of coarse particles will not ignite easily, and if ignited will burn rather slowly. However, if the given combustible material appears in the form of dust then its fire and explosion hazards are greatly increased. In fact, with dust, explosions, rather than fires, are more likely.

Dust particles in the range of a few tens of micrometers, dispersed in the air, behave in much the same way as a gas. Therefore, many fire and explosion characteristics describing hazards of a gas or vapor may equally be applied to dusts: dusts are explosive in a certain range of concentrations (lower and upper explosive limits), need a certain minimal energy of ignition, need a certain critical oxygen concentration for an explosion to occur, and may be rendered inert by an inert material. The destructiveness of a dust explosion can be described by the same factors which describe a gas explosion, such as the rate of pressure rise, maximum explosion pressure and the duration of excess pressure.

Besides these similarities there are some significant differences, however (Bardon & Fletcher 1983). The flame speed in the case of dust explosions is much slower (only up to 10 m/s), the rate of pressure rise is lower, ignition energies are higher (10–50 mJ, vs. 0.2–10 mJ for gases) and the explosive concentration range is much wider, with the upper explosive limits often several times the stoichiometric concentration. Another characteristic difference is the multiplicity of dust explosions (Palmer 1973), i.e. dust explosions usually occur in series. The primary explosion is often small and except for some minor material damage, no major structural and/or material destructions normally occur. Yet, the consequences of a dust explosion may be disastrous, as many recorded explosions show (see, for example, Corbert 1985). In virtually all these cases it was the second or even the third or fourth explosion that caused the major damage. The primary explosion disturbs settled dust in the vicinity and ignites the newly formed dust cloud, which in turn disturbs more settled dust and generates a still more serious explosion and so on.

The majority of dusts are generated, handled and stored in industrial settings. Technological operations that produce dust in the fine particle range are grinding, sieving, drying, conveying, collecting and spraying, while technological units most susceptible to dust explosions are dust collectors (cyclones, bag filters), storage vessels (silos, hoppers), pneumatic conveyors, grounding mills, etc.

Dust hazards can be conveniently described by an explosibility index, developed at the US Bureau of Mines. This index is defined as a product of the ignition sensitivity and the explosion severity of a dust. While the ignition sensitivity is primarily dependent on the ignition temperature and the minimal energy of ignition, the explosion severity is a function of maximal pressure and the rate of pressure rise. Dusts with an explosibility index below 0.1 are capable of producing only a weak explosion, while the explosion indexes 0.1–1.0, and 1.0–10 denote moderate and strong explosions, respectively. Dust explosions with an explosibility index above 10 cause severe material and structural damage. Some of the most hazardous dusts are listed in Table 2.6.

Field (1983) has given a small-scale explosibility assessment of industrial powders

Table 2.6 — Some of the most hazardous dusts (having an explosibility index greater than 10)

Type of dust	Material
Agricultural	Cellulose, alpha Cork Lycopodium Wheat starch
Coal	Lignite
Chemicals and drugs	Benzoic acid *Bis*-phenol A Hexamethylene tetramine Phthalic anhydride Aluminum tristearate Zinc stearate Sulfur Aspirin
Metals and alloys	Aluminum flake Magnesium, milled Titanium Uranium Uranium hydride Al–Mg alloy (Dowmetal)
Plastics and rubber	Acetal, linear Methyl methacrylate–ethyl acrylate copolymer Acrylonitrile polymer Cellulose acetate Nylon Phenol furfural Phenol formaldehyde Polyurethane foam Synthetic rubber

and dusts using the procedures of the Fire Research Station, England. The most significant parameters of dust hazards are: explosibility classification (explosive vs. nonexplosive dust), minimum ignition temperature, minimum explosible concentration, minimum ignition energy, limiting oxygen concentration to prevent ignition, maximum explosion pressure and maximum rate of pressure rise. The last two parameters were determined by both the Hartmann bomb and the 20 L sphere. Of 300 different dusts assessed, only 2% were found to be nonexplosive.

A critique of current experimental methods for the explosibility characterization of dusts (Kaufman *et al.* 1987) rightly states that these methods not only lack any fundamental character, but are also not able to predict secondary explosion hazards. Yet, as stated above, postprimary explosions of dusts are often the most damaging and hazardous.

2.1.5 Fire load as a measure of fire and explosion hazard

One of the principal combustion products is heat. The greater the heat liberated, the more intense is the fire severity. The most severe fires are the fully developed ones (see subsection 4.4.2). The maximum heat is released when all the combustibles, including interior finish, floor finish and structural elements in a given fire area, are burned. Obviously, the greater the mass of combustibles, m, and their heats of combustion, Q, the greater the maximum heat released. Qm is called the fire load. The theoretical fire load is always greater than the amount of heat actually released in a fire, since only rarely do ordinary combustibles burn completely.

As the mass of combustibles varies throughout a building, so does the fire load. Therefore, a related unit, termed the specific fire load, F, is introduced, which takes into account the unit area of the compartment, A_F, in which the combustibles are present. Since the mass of the material can be expressed as a product of its apparent density, ρ_{app}, and its volume, V, the specific fire load can be written as:

$$F = \frac{\rho_i V_i Q_i}{A_F} \text{ kJ/m}^2 \tag{2.8}$$

where Q_i is the heat of combustion of the ith material.

The values of the specific fire load are used for grading enclosures according to the fire hazard. Three groups of specific fire loads are usually defined, that is low, moderate and high. If the specific fire load of an enclosure does not exceed 1 GJ/m^2 of the net floor area, the enclosure is of low fire load. If the specific fire load exceeds 1 GJ/m^2, but is less than 2 GJ/m^2 of the net floor area, the enclosure is of moderate fire load. Finally, an enclosure containing combustibles in excess of 2 GJ/m^2 of the net floor area is of high fire load.

If the total area of the enclosure, A_F, containing combustible material and the value of the specific fire load, F_i, are known then the total fire load, G_o, within the enclosure can be written as:

$$G_o = F_i A_F \tag{2.9}$$

For comparative and practical purposes it is advantageous to express the mass of different combustibles as 'equivalent combustible weight', that is as the weight of ordinary combustibles having a heat of combustion of 18 600 kJ/kg which would liberate the same total heat as the given combustible. The 'equivalent combustible weight' is calculated as:

$$\text{Equivalent combustible weight} = \frac{\text{mass of combustible} \times \text{heat of combustion of given combustible}}{18\,600} \quad (2.10)$$

For example, the equivalent combustible weight of 25 kg of a metal with a heat of combustion of 12 070 J/kg is 16.2 kg of ordinary combustible.

The concepts of fire load and fire severity have been used for a long time to determine the fire resistance required for building partitions and structural elements, and to evaluate requirements for the basic elements of the active fire protection (the quantity of extinguishing agent, the location, number and equipment of fire fighting units, the arrangement and load of access lines and means of egress). However, fire evidence and research have shown that although the fire load is a measure of the fire and explosion hazard it cannot be accepted as the sole descriptor of the fire hazard. Other factors, such as the geometrical shape and size of combustibles, their arrangement, room geometry and size, are also important. Especially important is the amount of air available for combustion. Air (or more correctly, oxygen from air) takes part in the combustion process according to equation (2.2) on equal terms with the combustible material. That is, fire severity should be evaluated not only on the basis of fire load (that is, the 'concentration' of combustible material) but also on the basis of ventilation in fire (that is, the 'concentration' of oxygen in air) (see subsection 2.2.4).

2.2 SOURCES OF OXIDANTS

2.2.1 Oxygen in air

Referring to equation (2.2) it is clear that a fire or explosion can occur only if an oxidant is present together with a combustible material since only then, and in the presence of an ignition source, will the reductant be oxidized to the final combustion products. By far the most common oxidant is air itself, or, more correctly, oxygen in air, which comprises *ca.* 21 vol.% of the air–gas mixture. Oxygen in air comes in the form of molecular oxygen, O_2. It is a very reactive chemical species which is capable of oxidizing almost all other elements, or materials, producing oxides as the final reaction products. Other forms of oxygen, such as atomic oxygen, O, or ozone, O_3, are even more reactive. However, they are normally present in air only in negligible concentrations, leaving the molecular oxygen as the predominant oxygen species.

Compared to common oxidation reactions in which a material is slowly transformed to its oxidized form, fires and explosions can be regarded as oxidation reactions occurring very quickly. Of course, the role of the source of ignition is very important but only at the beginning of the reaction, that is up to the point when the overall reaction becomes exothermic. Since the reaction rate is controlled not only by the concentration of the combustible, but also by the concentration of air, it follows that all fire and explosion parameters, described earlier as the descriptors of fire and explosion hazards, reach their maximum in normal air, that is at the concentration of oxygen in air which is limited to 21% as the practical maximum. The remaining air constitutents, almost exclusively molecular nitrogen, do not take part in the oxidation reaction. Nitrogen, being an inert molecule, serves more as a moderator which controls the reaction rate. However, if for any reason the concentration of oxygen is changed, the reaction rate of the fire or explosion changes correspondingly. Thus, when the concentration of oxygen is gradually reduced and approaches zero, the overall fire and explosion hazard, and the parameters which describe it, are reduced. For example, ignition temperatures and energies become higher, the flammable range is narrowed, and the burning rate is decreased, leading eventually to termination of the fire or explosion.

On the other hand when the concentration of oxygen is increased, as in some oxygen-enriched atmospheres employed in some medical and industrial applications, the fire and explosion hazard in most cases increases considerably. For example, ignition temperatures and energies become lower, the flammable range is widened, the reaction rate is increased and flame resistance is less than in a normal atmosphere. Therefore work in oxygen-enriched atomspheres is generally more hazardous than when performed in air. In fact, many materials which in a normal atmosphere are considered noncombustible may burn, sometimes vigorously, in a pure oxygen atmosphere.

2.2.2 Chemically bound oxygen

Oxygen needed for the oxidation of the reductant (that is, combustible material) may also come from some compounds in which it is chemically bound in the form of various groups, such as peroxide O_2^{2-}, perchlorate, ClO_4^-, chlorate, ClO_3^-, chlorite, ClO_2^-, hypochlorite, ClO^-, nitrate, NO_3^-, nitrite, NO_2^-, permanganate, MnO_4^-, and persulfate, $S_2O_8^{2-}$. Most compounds containing these groups are not combustible themselves, yet they are considered hazardous because they can liberate oxygen needed for combustion (see section 2.1). In this way, the intensity of burning is increased since pockets of oxygen-enriched atmsopheres are formed in the immediate vicinity of the decomposed compound. Compounds containing bounded oxygen (oxidants) are widely used in many applications, such as salt baths, explosives, bleaching powders, fertilizers, etc.

The decomposition reactions leading to the release of oxygen:

$$O_2 \rightarrow O + 1/2\ O_2$$
$$ClO_4^- \rightarrow Cl^- + 2O_2$$
$$2ClO_3^- \rightarrow 2Cl^- + 3O_2$$
$$3ClO^- \rightarrow 2Cl^- + ClO_3^-$$
$$NO_3^- \rightarrow NO_2^- + 1/2\ O_2$$
$$NO_2^- \rightarrow NO + 1/2\ O_2 + e^-$$
$$3MnO_4^- \rightarrow 3MnO_2 + 3O_2 + 3e^-$$

occur mostly at higher temperatures, although some of the oxidants decompose at room temperature and/or in the presence of moisture. Several oxidants liberate oxidizing species other than oxygen, for example nitrous oxide, and chlorine.

If oxidants in finely divided form become intimately mixed with organic materials, explosive combustion can occur, initiated by a small ignition source, or in some cases spontaneously. Generally, chemically pure oxidants do not decompose readily, whereas contamination with combustible materials, or traces of metals and their salts promote decomposition.

The release of oxygen from oxidants involved in a fire is accompanied by the evolution of heat. When water is employed for extinguishing such fires, steam explosions are possible, further augmenting the hazard of oxidants.

2.2.3 Other oxidants

If fire is defined in more general terms as any exothermic reaction between an oxidant and a reductant, accompanied by the evolution of light and heated gaseous products, then the reaction of some materials with oxidants such as flourine, F_2, and chlorine, Cl_2, may be described as 'combustion'.

Fluorine is especially reactive. In fact, this is the most reactive of all chemical elements and forms fluorides with all ordinary combustibles at ordinary or elevated temperatures. Chlorine reacts similarly, although less vigorously.

Some metals, for example lithium, beryllium, calcium, magnesium, aluminum and many transition metals, react at moderate or slightly elevated temperatures with nitrogen, giving corresponding nitrides (M_3N_2, or MN, depending on the valence of the metal). Reactions of Li and Mg with nitrogen are quantitative so that these metals can be used to remove nitrogen from other gases. Similarly, some metals can burn in the atmosphere of carbon dioxide, forming corresponding carbides as products of reaction.

2.2.4 Ventilation as a measure of fire hazard

According to equation (2.2) the intensity of a fire or explosion is determined by both the concentration of fuel (fire load, subsection 2.1.4) and the concentration of air. The well known Le Chatelier's principle in chemistry governing the rate of a chemical reaction states that the reaction rate is determined by the concentration ratio of the reactants. When one of the rectants is in excess then the other one controls the reaction rate. This rationale can be applied as well to the description of fire and explosion hazards. The dependence of a fire hazard on the concentration of fuel (fire load) has been discussed in subsection 2.1.4, whereas the dependence of a fire hazard on the concentration of air will be briefly outlined here. A more detailed discussion is given in subsection 4.4.2.

The hazards of a fire occurring in the open are practically independent of the concentration of air, that is on the ventilation, since the fuel is always the limiting reactant because of the enormous excess of air. In other words, such fires are always fuel (or, more correctly, fuel-free surface area) controlled (see section 4.4). On the other hand, fires in enclosures can be, and often are, ventilation controlled. During the pre-flashover period of fire growth in an enclosure, the fire behavior (and hazards) is similar to that as if the fire were developing in the open since there is an ample amount of air available for oxidation of the fuel. However, after the flashover period, that is during the fully developed fire, the principal descriptors of fire hazards, such as the rate of burning, the rate of heat release, the duration of fire, and the maximum temperature, are strongly dependent on the ratio of air to fuel. At some optimum ratio of these two reactants, termed 'critical', fire hazards are approximately equally dependent on concentration of both the fuel and the air. If the supply of air is less than that determined by the critical ratio, fire hazards are controlled by the ventilation, that is the rate of air flow. Similarly, if the ventilation is such that the supply of air is greater than that determined by the critical ratio, fire hazards will be controlled basically by fire load. Quantitative relationships between some of the descriptors of fire hazards and ventilation are given in section 4.4, describing the dynamics of fires in enclosures.

2.3 SOURCES OF HEAT ENERGY

The third factor in equations (2.1) and (2.2), a source of heat energy necessary for the initiation of the reaction between a fuel and an oxidant, is indeed the most important one. Only in some exceptionally rare circumstances will the sole contact between the fuel and oxygen lead to a spontaneous combustion, or explosion. In the great majority of cases an outside source of heat energy is necessary for the initiation of the combustion. The source of heat energy often behaves as a catalyst, especially in the piloted ignition. Its presence is necessary for the promotion of the combustion reaction, but only up to the point when the reaction becomes self-sustaining.

There are several distinct means of generating heat energy in the quantities and duration required for the initiation of a fire or explosion. They may conveniently be divided into three broad categories, such as electrical, chemical and mechanical sources of heat energy. The hazards associated with each of these sources are in many ways specific and therefore deserve separate consideration. Yet all three sources of heat energy have one common and most important characteristic: if uncontrolled, each of them will cause a fire and/or explosion, given that the other reactants in equation (2.2) are simultaneously present.

2.3.1 Electrical heat energy

Electricity is repeatedly quoted as a major cause of many fires and explosions. Fire statistics regularly show electricity to be responsible for between 20 and 40% of all fires and explosions of known origin. Although formally it is true that electrical heat energy does cause a significant number of fires, it is also indisputable that fires will seldom, if ever, be initiated by the electricity if the electrical installations and equipment are made, used, and maintained in accordance with the corresponding codes and standards.

Electrical heat energy is always generated when an electric conductor carries a current. Joule's heat formed, Q, is directly proportional to the square of the current, I, the resistance of the conductor, R, and the time the current flows through the conductor, t,:

$$Q=I^2Rt \qquad (2.11)$$

Obviously, the greater the current, the resistance and the longer the current flows through the conductor, the greater the heat energy liberated. For safe use of electrical conductors, electrical codes specify the maximum current which will not overheat them. The resistance of conductors is determined by the kind of material employed. Materials which are good conductors, such as copper, silver, and aluminium, are preferred and generally used. Finally, the time the current takes to pass through the conductor is determined by the intended use of electrical appliances and equipment, unless the appliance is left on unintentionally, causing an accidental fire or explosion.

Since electrical conductors are shielded by a layer of insulating material and a protective cover, the heat liberated under ordinary circumstances is so slight as to be negligible. The question may be asked: why, then, do electrically caused fires and explosions occur? The answers are many. For example, there may be few circuits, and these may be of inadequate capacity, thus raising the values of I and Q in equation (2.11). Or, the overcurrent protective devices, such as circuit breakers and fuses, may be of greater size than required for the circuits they protect, again leading to overheating of the circuit. Further, the circuit may be interrupted on purpose or unintentionally. While the intentional circuit interruption is safely regulated, for example by a switch, the second one, caused, for example, by accidental cutting, loosened terminals or defective installation invariably leads to the release of electrical heat in the form of arcing. The energy of the arc is determined by the current strength, the voltage and the resistance. The temperature of the arc may easily reach 3000–5000 K. Whether the insulation and the conductor cover, the metal from which the conductor is made or the nearby combustibles will catch fire depends on the duration of the arcing and the time necessary for the combustibles to be readied for burning. When the duration of the arcing is longer than the latter time, a fire or explosion will ensue. Similarly, if the conductor insulation is worn out or damaged, the Joule's heat may be high enough to ignite the insulation or other combustibles.

In locations where flammable liquids, gases or dusts are produced, stored, or handled, special electrical installations and equipment must be used as laid down by the requirements and specification set forth in the codes and standards. As to the number of regulations relating to the electrical installations and equipment, it is felt that this field, although indisputably important from a fire and explosion point of view, is really overregulated internationally, regionally and even nationally (Short 1987).

Electrical conductors, installations and equipment in fire conditions pose a serious hazard to life. In addition to the well known hazard to life of high voltages people and materials may be threatened by toxic and corrosive products released during the combustion of the conductor insulation and cover. For a long time, PVC-based insulators have been extensively used, since this material possesses many good and desirable characteristics. However, PVC in a fire atmosphere, although practically incombustible, releases large quantities of HCl gas (roughly 0.3 kg HCl per kg PVC) which readily combines with water to form a strong and corrosive hydrochloric acid. Especially hazardous are fires in electric transformers. The fire hazard of this equipment is determined by the fire characteristics of liquids used for cooling and by the insulating materials (Altman & Pfeiffer 1984). Up to the 1980s the predominant liquid used in transformers was a mixture of chlorinated biphyls and chlorobenzene, known as askarel. It is electrically stable and is not flammable. However, in some countries, concern for environmental protection has been repeatedly expressed, since askarels in fire release some compounds (such as dibenzene dioxin and furans) which are proven cancerous chemicals. Therefore in a number of countries, askarel-containing transformers have been removed and replaced with environmentally acceptable cooling liquids, such as mineral or silicon oils or pastes.

Electrical heat energy as a fire hazard may originate from natural sources also. The discharge of atmospheric electricity, known as lightning, has been a constant cause of fires. Around 300 fires caused by lightning are recorded annually in the UK alone [Simpson *et al.* 1985). Although prevention measures have decreased the number of fires and explosions, lightning still causes considerable damage to high buildings and industrial plants, owing to high energetic parameters: the electric current is normally between 10 and 100 kA, while the heat released may reach 1000 MJ. Even though the discharge time is measured in milliseconds, its duration is long enough for many combustible materials to be ignited.

Another natural source of electricity which may present a fire or explosion hazard is static electricity. It is formed whenever materials of dissimilar conducting properties are in contact. The accumulation of static electricity is not itself dangerous. What is hazardous is the formation of small sparks as a natural means of bridging oppositely charged surfaces. The energy of such sparks, though fairly low (often a few mJ), may be high enough to ignite a flammable mixture.

Fire hazards of electrical installations and equipment are not only caused by overheating and/or arcing, but also by nonelectrical means. For example, when an electric air heater is in direct contact with a combustible material the release of heat is blocked or diminished, leading to hazardous conditions. Or, when an electric motor is left unlubricated from some time, motor bearings may overheat, causing fire or explosion, with mechanical, rather electrical, heat energy as a source of heat energy.

2.3.2 Chemical heat energy

Heat energy necessary for the initiation of fires and/or explosions may be produced by several types of chemical reactions. Among them of primary concern to fire and explosion protection are combustion reactions. These reactions when brought to a completion release a considerable amount of heat, known as the heat of combustion (see subsection 2.1.1). Examples of common sources of heat energy formed by combustion reactions are open flames, a lighted cigarette, an acetylene torch used for cutting and welding, a hot fuel-fired boiler, a drier or a furnace surface.

The heat of combustion is determined calorimetrically. It varies with the chemical composition, structure and aggregate state of the material and is expressed in megajoules per kilogram of material. Heats of combustion for pure gaseous hydrocarbons are between 46 and 55 MJ/kg. The introduction of one or more oxygen, nitrogen, sulfur or halogen atoms into the hydrocarbon molecule considerably lowers the heat of combustion. Heats of combustion of solids and liquids are generally lower than those of gases owing to the latent heat of vaporization. Most of the common combustible materials, such as coal, coke, textiles, common plastics, gasoline, fuel oil, wood, etc., release 15–45 MJ of heat per kg of fuel. In practice, combustion is frequently not completed owing to the lack of air in the enclosure where the fire is taking place. Therefore it is the oxygen in the air which often determines the amount of heat released in a combustion reaction (see subsection 2.1.1).

Other types of chemical reaction which release heat are spontaneous heating (or self-heating) and spontaneous ignition (self-ignition). As the name implies, the initiation of heating and ignition in such reactions does not require an outside source of ignition. All organic substances capable of combustion are incessantly oxidized by oxygen from the air, even at subnormal temperatures. However, the rate of oxidation is normally too slow to lead to a perceptible heating and ignition, since the minute amounts of heat released in such an oxidation are transferred to the environment as soon as they are formed. If the dissipation of heat is blocked for any reason (for example, by the interplay of factors such as air supply and insulation) the heat released may accumulate and eventually lead to self-heating and self-ignition. Although, theoretically, many organic materials should be capable of spontaneous heating and ignition, few of them actually do so. Thus, cases of spontaneous heating and ignition of coal, especially freshly mined, and of some agricultural products are recorded each year in many countries. According to most fire statistics spontaneous heating is the source of ignition for only 0.5–1.0% of all fires, with a somewhat larger percentage in the chemical industry.

Even rarer are cases of ignition due to the heat of dissolution of a substance. Although the dissolution of the majority of materials in a liquid is accompanied by the release of heat, the amount of heat is normally too low to be capable of igniting the material itself or the nearby combustibles. As notable exceptions the dissolution of concentrated sulfuric acid or solid sodium hydroxide in small volumes of water may be cited.

2.3.3 Mechanical heat energy

It is well known that some forms of mechanical energy can be converted to heat energy. The greater the mechanical work, the greater is the amount of heat released. The mechanical heat energy, formed as a result of mechanical failure or malfunction of the processing equipment, is the source of ignition for a significant number of fires (5–10%), while in coal mines it is the leading source of methane explosions.

When two moving solid surfaces are in contact (for example a belt and roller(s) on a conveyer carrying a combustible material) the resistance to relative motion is manifested in the form of heat due to friction. If heat due to friction is not dissipated as rapidly as it is formed there will be overheating which may ignite the combustible material. If one of the solids is metal, frictional heat may appear in the form of sparks. The energy content of the spark depends on factors such as the heat of combustion of the metal employed and the size of the particle carrying the heat. Small particles of metal have a small total heat content and may not be capable of igniting the combustibles though their temperature may be over 1000 K. On the other hand, larger particles possess a greater heat content, but are of lower temperature, owing to their larger size.

Mechanical heat energy may also be generated when a gas is compressed. The resulting heat of compression may be sufficiently large to ignite a flammable gas or vapor. Practical use of this form of mechanical heat energy is to be found in diesel engines. However, there are comparatively few instances where a serious fire or explosion has been reported which was caused by the heat of compression of air or some other gas.

2.4 SOURCES OF EXPLOSION HAZARDS

To most people, explosions — rather than fires — are events that have far greater potential for injury to human beings and for damage to property. Fire and explosion statistics disclose that although explosions are generally fewer in number annually, they do cause comparatively greater damage. Accordingly, a separate subsection on the sources of explosion hazard is warranted.

There are many definitions of explosion in the literature. Marshall (1987) has compared eight references which give definitions of an explosion, and noted that most of them involve such concepts as the suddenness of the event, the release of internal energy, the pressure rise and the appearnce of a loud bang or noise. He offered a definition which says that 'an explosion is a sudden release of energy accompanied by the generation of a pressure wave and by a loud noise. It has a potential for injury to human beings and for damage to property. Explosions may be physical or chemical . . . '.

A graphical presentation of Marshall's classification is shown in Fig. 2.3. In

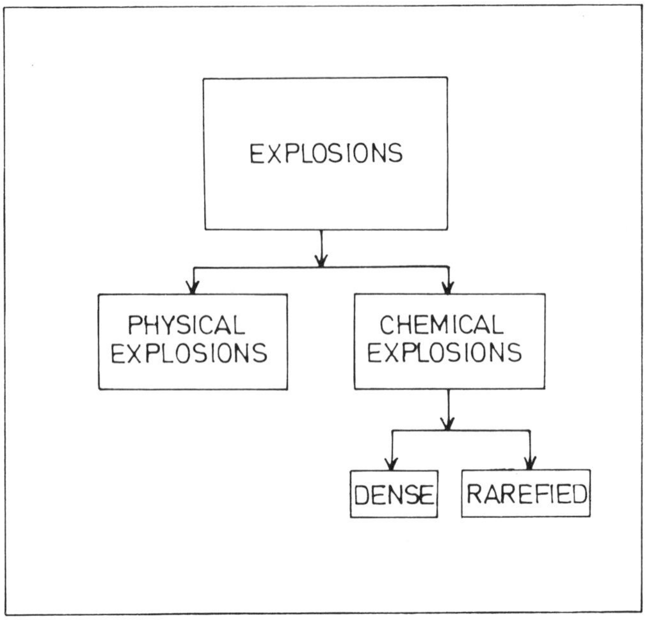

Fig. 2.3 — Classification of explosions.

keeping with the concept of this book, which defines fires and explosions as purely chemical events definable by equation (2.2), physical explosions, although capable of exerting substantial damage, are of no concern here. Nor are explosions caused by conventional (also known as 'dense') explosives, since hazards of these important industrial and military products are well known and defined in regulations underlying their production, transport and use. Instead, hazards of explosions originating as a result of chemical transformations of ordinary materials will be discussed in some detail.

As previously stated, chemical explosions may be defined by the same basic approach (equation (2.2)) which also holds true for fires. It means that for an explosion to occur, the same requirements, such as the presence of a combustible material, of an oxidant and of a source of heat energy, must be satisfied. A principal difference between fires and explosions lies in the reaction rate. The rates of chemical reactions ending in an explosion are of the order of milliseconds for deflagration reactions, or even shorter for detonations. Another difference is the absence of a pressure wave and a loud noise in the case of fire. These differences can be explained on the basis of different initial conditions before the ignition. That is to say, explosions generally occur in homogeneous mixtures of fuel and oxidant, so the burning rates are not restricted by the supply rate of fuel and oxidant, whereas the reaction rate of chemical reactions characterized as a fire is determined primarily by the rate of supply of fuel and oxidant to the flame. Sometimes the products of explosion combustion reaction are different from fires, since besides CO_2 and H_2O many other gaseous compounds in substantial concentration may be formed as principal combustion products. Nevertheless, hazards of an explosion can be conveniently classified in the same manner as the hazards of a fire. That is, in order to undertake optimal prevention measures, one must known the sources of the explosion hazards pertaining to the combustible material, to the oxidant and to the source of heat energy, respectively. Since these factors have already been discussed in detail elsewhere (see sections 2.1., 2.2 and 2.3), the following lines will emphasize the explosibility aspects only.

For a chemical explosion reaction to occur, the material must be combustible. Of the three different aggregate states, the materials most susceptible to explosions are the gaseous ones. When they are intimately mixed in the right proportions with air, the presence of an ignition source invariably causes an explosion. The concept of the lower and upper explosibility limits, and the influence of the temperature and/or pressure of the mixture on the explosibility limits have already been discussed (see sections 2.2, 2.3 and 4.6). The power density, or energy released in unit time per unit volume (Marshall 1987), for gas and vapor explosions is substantially lower than for dense explosives. The vast majority of gas explosions are deflagrations. Gas detonations are possible only in instances when the ignition source is of high energy, or when the gas is confined in certain geometries, such as long pipes, or long underground galleries. Since gases and vapors are widely used, it follows that hazards of gas and/or vapor explosions are present in many public sectors and in various industries. From a practical point of view it can be said that whenever a flammable gas is released in a confined space there is an explosion hazard. Gases are explosion hazards even when unconfined, but only in very large quantities, exceeding at least three tonnes (Marshall 1987). Confined gas explosions causing great losses frequently occur in coal mines, in the gas and chemical industries, in storage tanks, in cargo spaces of supertankers, in industries using spray finishing and dipping and coating processes, in boiler furnaces, etc.

Other aggregate states of a material may also be explosion hazardous. For example, solid combustible materials in the form of a dust, or a powder, with particle size less than *ca.* 840 μ, behave in a similar manner to gases. So do drops of combustible liquids dispersed in air in the form of a mist. However, the explosion characteristics of such dispersions are highly subject to turbulence effects which generally increase the power of an explosion. Among the industries possessing a high hazard of dust explosions are coal mining, the grain industry and facilities handling bulk raw agricultural commodities, the processing and finishing of metals (Al, Ti, Zr, Mg), and the wood-working and wood-processing industries.

Most combustible solid materials, unless in a dust or powder form, are not explosion hazardous under ordinary circumstances. Exceptions are some solid materials used in the chemical process industries in reactions such as nitration, polymerization, diazotization, decomposition and condensation. These reactions, if uncontrolled, tend to run away, with reaction rates accelerated to such a degree that deflagration is often transformed into detonation.

3

Control of fire and explosion hazard sources

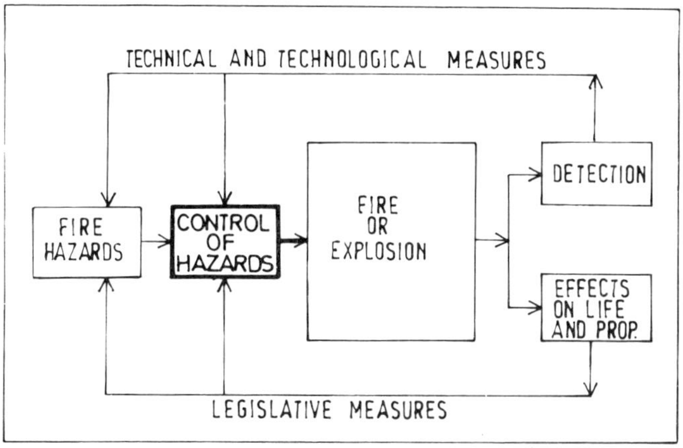

This block in the scheme of the system approach to fire and explosion protection is connected to both the block describing the character of various fire and explosion hazard sources (Chapter 2) and the block dealing with the dynamics of fires and explosions (Chapter 4). Once the nature and characteristics of all sources of fire and explosion hazards are known, the next logical step to prevent, or to extinguish, a fire or explosion is to take measures to control hazard sources. The various kinds of measures may be divided into technical and technological, and legislative measures. These measures must act upon not only the control of hazard sources but also on the hazard sources themselves. The ultimate goal of fire and explosion protection is to prevent fires and explosions, that is to terminate the protection systems at the block pertaining to the control of hazard sources. However, should a fire or explosion break out despite the protective measures applied at the hazard sources and their control, then the next obvious goal is to strengthen the quality and duration of the control measures to the point that the dynamics of fire and explosion are such that none or only negligibly harmful effects are manifested on people and property.

This chapter is concerned only with the theoretical basis of control of sources of fire and explosion hazards. Numerous practical means of realizing the theoretical postulations are dealt with in Chapters 7 and 8.

Throughout this book, fires and explosions revolve around the basic equation (for example, equations (2.1), (2.2), (4.1), (4.2), (6.1)) describing the necessary requirements to be fulfilled for the chemical reaction between a combustible material and an oxidant initiated by a source of heat energy. The equation is shown once more (equation (3.1)) since it represents a basis for the effective design and application of various control strategies:

$$\boxed{\text{Fuel}} + \boxed{\text{Air}} \xrightarrow[\Delta]{} \boxed{\text{FIRE or EXPLOSION}} \qquad (3.1)$$

The hazards of fuel, of air (as the source of oxidant), and of three principal means of generating heat energy are described in Chapter 2. Efforts to control hazard sources are aimed at either preventing initiation of the reaction represented by equation (3.1), or its termination, if the combustion reaction has already begun. Accordingly, the principles of achieving this object can be classified into two broad groups, the preventive and the protective.

3.1 PRINCIPLES OF FIRE AND EXPLOSION PREVENTION CONTROL

The principle of *preventive control* of hazard sources, or, generally, of preventive fire and explosion protection is fairly simple: every effort should be directed toward the avoidance of a simultaneous presence in time and space of fuel, air and a source of heat energy in concentrations or intensity leading to the commencement of the chemical reaction shown in equation (2.1). This means that when at least one of these three ingredients is absent, or is present only in a low concentration or intensity, there will not be a fire or explosion. Graphical representation of this statement can be shown by equations (3.2), (3.3), and (3.4):

The prevention of fires and explosions by control of the fuel:

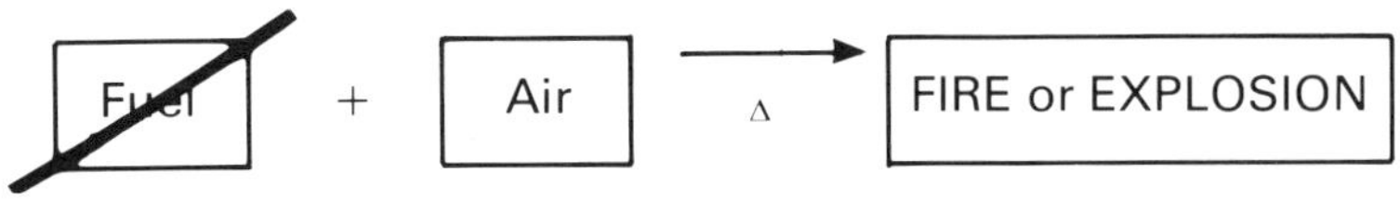

(3.2)

The prevention of fires and explosions by control of the oxidant:

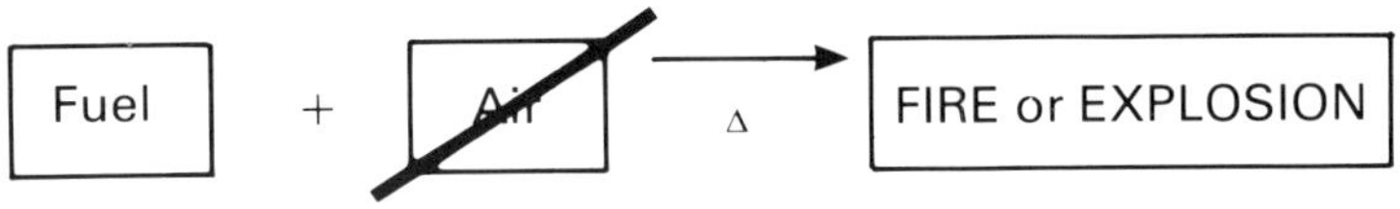

(3.3)

The prevention of fires and explosions by control of the sources of heat energy:

(3.4)

Theoretically the best control of fuel as a means of preventing fire and explosion could be achieved if it were completely absent, removed or physically separated from the other reactants in equation (3.1). In practice control of the fuel is often the least practical strategy to prevent fire and explosion, since a completely fuelless environment is almost impossible. For example, the usual content of a building, of an apartment, or of an economically sound technological process dealing with flammable or combustible materials cannot be changed for the sake of fire and explosion prevention alone. Nor is it always possible to replace hazardous materials with less hazardous, or with incombustible ones. It will be shown, however, that means of controlling some of the principal fire and explosion characteristics of fuel such as flame spread, rate of heat release, smoke production, toxicity, ignition temperature, burning rate, flammability (explosibility) limits, etc., do exist and may be used as a complement to the other fire and explosion prevention measures.

The other possible means of fire and explosion prevention is control of the oxidant, as the second reactant in equation (3.1). This control can be achieved by several modes. For example, the oxidant may be removed altogether, as in some highly hazardous chemical processes where the use of vacuum or a completely inert atmosphere is required. In other cases the concentraion of oxygen in air, as the most ubiquitous oxidant, may be decreased by dilution with an inert gas from the normal 21 vol.% to the point that the atmosphere no longer supports the ignition and combustion of fuel. Yet another means is the removal or separation of the sources of oxidant from the fuel if they are in the form of bounded oxygen, as is oxygen in the oxidizing chemicals. While control of the oxidant is a proven means of prevention of fire and explosion in some rather specialized cases, the fact is that in the great majority of normal situations the control of air, just like the control of fuel, is impractical or unfeasible.

This leaves the control of sources of heat energy necessary for the ignition of fuel as the most important means of fire and explosion prevention. Indeed, except in rather rare cases of self-ignition, no fire or explosion will ever commence by the reaction of fuel with air alone. A heat energy source of such duration and intensity as to be capable of raising the temperature of the fuel to the ignition point is always the main prerequisite for fire and explosion. The three principal sources of heat energy (electrical, chemical and mechanical) are intimately associated with almost every aspect of human activity. For example, electricity is the backbone of industrialized societies, while combustion processes are a basis for such irreplacable needs as heating, cooking and transportation. Thus, it is indeed a big challenge to keep all the numerous forms of sources of heat energy within the safety limits. If arson and natural phenomena are excluded, the other known causes and ignition and fire spread are all attributable to one or more possible sources of heat energy. Yet, this is maybe too harsh a statement, unjustly condemning sources of heat energy themselves. The fact is that human failure of one or another kind is the most frequent cause of fires and explosions since it is through human negligence, failure or purpose that the ignition heat and ignited material are brought together. Even equipment failure as a cause of fire and explosion can for the most part be blamed on people rather than on the equipment.

In reality, the control of heat energy is the most important for preventing fires and explosions. Electrical sources of heat energy should be controlled in such a way that sparking, arcing, resistance heating or static electricity are avoided, or, if unavoidable, are incapable of raising the temperature of the surrounding combustibles above the ignition point. Sources of chemical heat energy — for example, open flames, smoking, matches, heating and cooking equipment — should be kept under control at all times since when improperly used, maintained or operated they are responsible for a large proportion of fires and explosions, as are sparks, overheating of machinery and friction as the principal sources of mechanical heat energy.

It should be borne in mind that although the control of one of the reactants in equation (3.1) is a proven means of fire and explosion prevention, the best prevention is accomplished when more than one of the reactants are controlled. For example, it would be unwise to base explosion prevention in a flammable air mixture only on the control of the electrical equipment, when ventilation, which prevents the formation and accumulation of the explosive mixture, can additionally tighten control.

3.2 PRINCIPLES OF FIRE AND EXPLOSION PROTECTION CONTROL

The *protective control* of hazard sources, also known as repressive control, fire extinguishment, or fire management control, is based basically on the same principles as preventive control. The main difference is that protective control is applied after a fire or explosion has already started when all efforts should be concentrated on extinguishing the fire or explosion. Another important difference is in the nature of the initiation source. While a fire or explosion can be initiated by a number of well characterized sources of heat energy, its further growth does not necessarily require the presence of the original ignition source. Fire flame and heat, as the principal combustion products, become promoters of further combustion reaction. Thus the original source of heat energy can be regarded, in a way, as a catalyst for the combustion reaction. Using the same graphical presentation by means of block equations, the essence of fire protection can be summarized by equation (3.5):

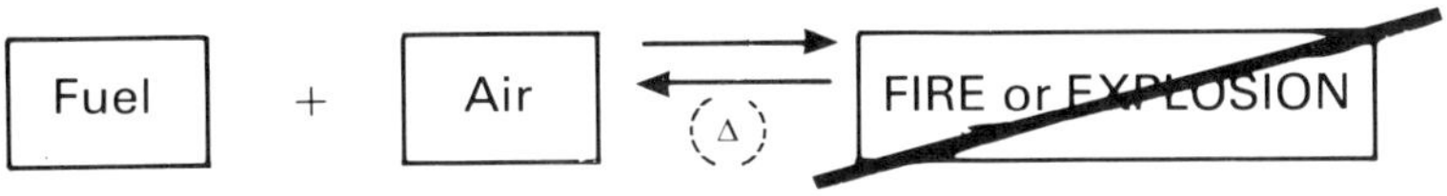

(3.5)

For the purpose of emphasizing the important means of achieving protection, the reaction (3.5) has been written as a reversible one. This means that even though control of the right side of equation (3.5) is the principal mode of fire and explosion protection, the combustion reaction can be reverted (and the fire and explosion thus extinguished) by controlling the factors on the left side. Note that the source of heat energy is now put in dashed brackets, implying that a developed fire or explosion can occur without the original ignition source. Again, it is important to emphasize that in practice, fire and explosion protection can be performed by a sumultaneous control of more than one factor in equation (3.1).

Regardless of the mode in which a fire or explosion occurs (flaming, nonflaming), principal attempts in fighting a fire are directed at removing heat from both the fuel under fire and the unburned fuel. The aim is to lower the temperature of the fuel below the ignition temperature. The removal of heat can be affected by cooling with an extinguishing agent, or by inhibition of the flames by chemical means. The most effective cooling agent is plain water, or water treated with some agents which make cooling more effective. Flames can be inhibited by a number of chemicals added to the flames. The inhibition is based on still poorly understood chemical reactions which effectively stop the flames and thus extinguish the fire and explosion. The most efficient flame inhibition agents are certain halogenated hydrocarbons and some inorganic salts. Smoke and gases as the principal products of combustion in enclosures can be controlled by an orderly removal from the fire scene, while excessive pressure, accompanying explosions, can be reduced by the proper application of various venting procedures.

Fire and explosion extinguishment can also be successfully accomplished by controlling the air (oxygen dilution) and the fuel. In such cases the reaction (3.5) is reversed, since one (or both) of the reactants is being continuously removed, thus stopping the combustion reaction. The theoretical principles of fire and explosion control by the control of fuel and air have been described above in the discussion on principles of fire prevention.

4

Fire and explosion dynamics

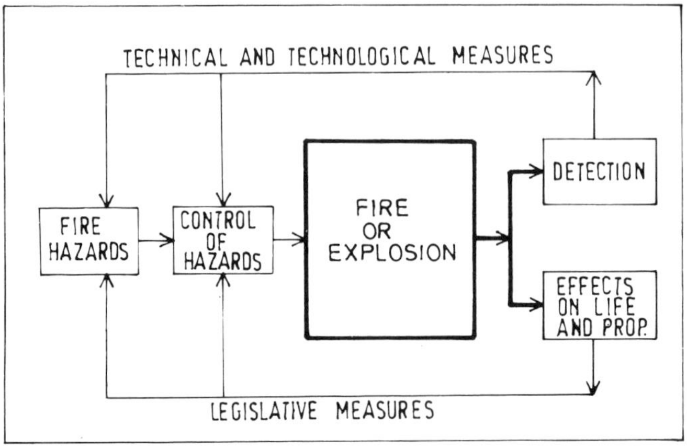

Regardless of the kind and quality of prevention measures applied, fires and explosions, being probabilistic events, do occur, and exert negative effects on people's lives and material goods. However, once a fire has broken out, there are a number of questions to be answered, as for example:

— will it die out or will it grow or spread?
— what are the controlling factors?
— how long will it last, and what intensity will it attain?
— what size will it be?
— what will be the temperature of the flames?

— what would be the quantity of smoke developed during the burning, its composition and modes of its dispersion?
— what volume of fire gases will be liberated?

Of course, the number of possible questions is almost unlimited, but answers to them are as yet few, and are often only empirical values. Fire science, being still in the developing stage, has not yet progressed to the levels achieved by the more traditional fundamental and applied sciences. One of the most important reasons is that a fire or explosion process is of such great complexity, resulting from the interaction of a large number of chemical and physical reactions and processes, that it is extremely hard to find a satisfactory answer to the above questions. However, knowledge of some basic fire and explosion components is steadily improving, enabling one to give fairly correct answers at least to those questions relating to human safety and the reduction of material and property losses.

In the systematic consideration of fire and explosion protection, knowledge of fire and explosion dynamics is important in relation to the other components of the system, especially the ones that refer to the detection, and the effects of fires and explosions on people and the environment (see section 1.4). Obvious goals of fire and explosion protection are the early detection of these events and the least harm to people and damage to materials and goods as is possible. These goals will certainly be helped if the mechanisms of fires and explosions from the incipient stage to their spread, to the unmanageable phase, are understood.

The spatial and temporal growth of fire, that is, its dynamics, depends to a considerable degree on whether it takes place in the open or in an enclosure. Therefore it is useful to classify fires into two broad categories, each of which may be subdivided into additional classes, as shown in the following section.

4.1 CLASSIFICATION OF FIRES

Depending on the condition of gas and heat exchange with the surroundings, all fires can be divided into two broad categories: fires in the open and fires in enclosures (Kimstach *et al.* 1984). Taking the rate of fire spread and the rate of air inflow as other criteria, further subdivisions can be made (Table 4.1). Thus, fires in the open

Table 4.1 — Classification of fires

Types of fires	
Fires in the open	Fires in enclosures
Localized fires	
	Fires in ‘open’ enclosures
Spreading fires	
	Fires in ‘closed’ enclosures
Massive fires	

may be of three types: localized, growing, and massive, while fires in enclosures may be divided into two subclasses, depending on which of the reactants in the basic fire equation (4.1) controls the reaction rate:

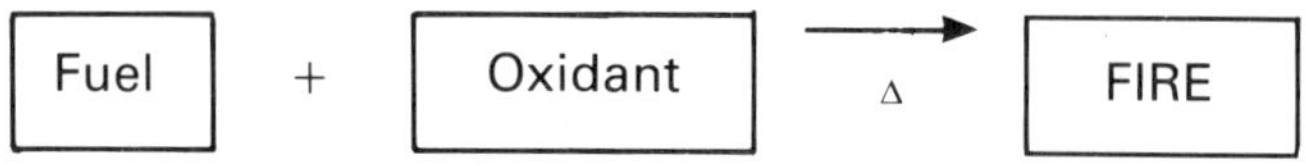

(4.1)

Ideally, fuel as one of the reactants is oxidized stoichiometrically with air as the source of oxidant (designated as boxes of the same height in equation (4.1)). However, if there is ample air around the fuel (shown as the larger size oxidant box in equation (4.2)), or, if the fuel supply is rather limited,

(4.2)

then, according to Le Chatelier's principle, the rate of the above equation is controlled by the rate of fuel supplied. In other words, such a fire is fuel controlled. Conversely, if the concentration of fuel is prevailing and air supply is scant, fire is ventilation controlled. Thus, fires in enclosures can be divided into fuel controlled (in 'open' enclosures) and ventilation controlled (in 'closed' enclosures).

Localized fires

These are confined to a restricted area, with no further spreading. An example of a localized fire in the open is a fire on an electrical transformer (Fig. 4.1(a)). Of course, before becoming localized, a fire has to spread a certain distance, determined by the availability of fuel, so localized fires might be regarded as a special case of spreading fires.

Under adverse meteorological conditions, such as strong winds directed to nearby combustible structures or materials, convective or radiative heat transfer or flying brands from localized fires may transform them to spreading fires, or even to massive fires.

Spreading fires

If a fire breaks out in a combustible environment it will spread and grow in size until either all the fuel is consumed or the fire is extinguished. The distance and the rate at which a fire will spread are often unpredictable. Predominant directions of fire spread are determined by the arrangement of combustibles on the given area and by environmental factors, such as wind, surface moisture, etc. (Fig. 4.1(b)).

Massive fires

Massive fires are really spreading fires occurring over a large area, where many relatively small spreading and localized fires all coalesce into a gigantic fire (Fig.

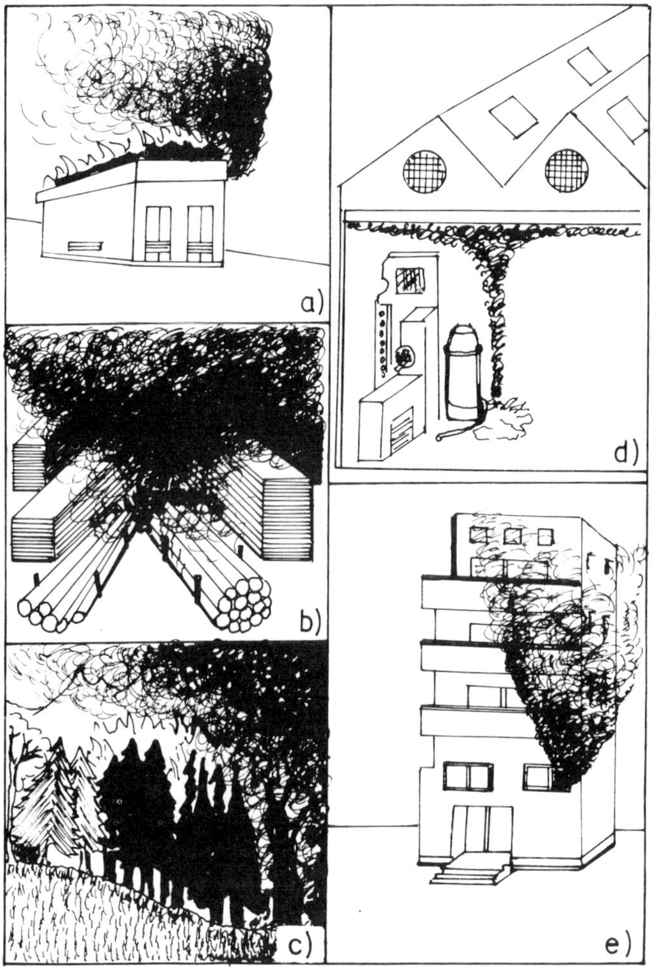

Fig. 4.1 — Classification of fires: (a) localized, (b) spreading, (c) massive, (d) in 'open' enclosure, (e) in 'closed' enclosure.

4.1(c)). Examples of such fires are forest fires which may encompass huge areas, e.g. several hundreds, or even thousands, of square kilometers, or fires in towns built mostly of wooden houses (also called 'conflagrant' fires, or 'conflagrations').

Massive fires are difficult to control and extinguish. Heavy rains in the stricken region are frequently the only way of stopping and extinguishing them.

Fires in 'open' enclosures

These fires occur in large confined spaces, with high ceilings and large openings, where plenty of available air makes them resemble fires in the open (Fig. 4.1(d)). They are characterized by high burning rates, but are not highly destructive. Physicochemical characteristics of combustibles, their spatial arrangement and quantity in the enclosure control the fire duration. Flames usually emerge from any openings and may spread fire to combustibles outside.

Fires in 'closed' enclosures

In rooms or other spaces with small openings, or without windows or doors, in the absence of forced movement of air, exchange of air and gases is limited. Such conditions prevail in building basements, garages, night clubs, etc. Fires in such environments (Fig. 4.1(e)) are therefore ventilation controlled, usually last longer and may be highly destructive owing to high temperatures in the period of full development. The rate of burning does not depend so much on the fire load as on the rate of air inflow. Copious amounts of smoke and toxic gases produced during the burning make ventilation-controlled fires very dangerous to people.

4.2 IGNITION AND SPREAD OF FIRE

Fire will start when a strong ignition source is in contact with, or in close proximity to, a flammable or combustible material for a time long enough to establish a self-sustained combustion. This outline will deal with the ignition of solid materials, but the principles are valid for the description of ignition of liquids and gases since the essence of ignition in most cases is the formation of combustible vapors from the material and its reaction with an oxidant. In the case of liquids, and especially gases, flammable vapors are already present, even at ordinary temperatures, which means that they are generally much easier to ignite than solids.

The mechanism of ignition is fairly well understood and there are a number of good reviews which cover it (Emmons & Atrega 1982, Thomas 1975). Although there might be some variation in the details, it is generally agreed that the following steps are essential for ignition of a material to occur.

Except for a rather limited number of materials exhibiting self-heating, the heat needed to sustain the combustion process must be supplied from the outside (Fig. 4.2(a)) for a sufficient time, until a critical heat transfer is reached (Rasbash *et al.* 1986). Physical and chemical characteristics of the material determine to a large extent the ease or difficulty of ignition. Among them the most important are those that define 'thermal inertia', that is, the thermal conductivity, K, the density, ρ, and the specific heat, C, Harmathy (1985) has proposed using the term 'thermal absorptivity', defined as $(K\rho C)^{1/2}$, as the characteristic of material in a fire environment.

The thermal conductivity determines the degree of insulation of the surface. If K is low, the material is a good insulator and its surface heats up rather quickly, and vice versa. Similarly, if the density of the material is low, its surface heats up faster than if the material is of high density. The effect of C is less pronounced, but generally good insulators have higher values of C. The thermal absorptivity $(K\rho C)^{1/2}$ is therefore a convenient criterion for the ease with which a material may be ignited. The lower the thermal absorptivity of a material, the lower the loss of heat by conduction through the material and the higher its surface temperature, hence the easier its ignition (Fig. 4.3). The product $K\rho C$ $(\mathrm{W\,m^{-2}\,K^{-1}})^2\,\mathrm{s}$ ranges from *ca.* $1.9–6.2\times10^3$ for polyurethane foam, to 1.98×10^5 for wood, to 1.3×10^9 for copper (Wooley 1985).

Prolonged heating of the surface of the material leads to a series of complex chemical reactions termed pyrolysis which decompose the material into volatile compounds of a lower molecular weight and a solid residue (char, in the case of cellulosic materials) (Fig. 4.2(b)). This process does not necessarily require the

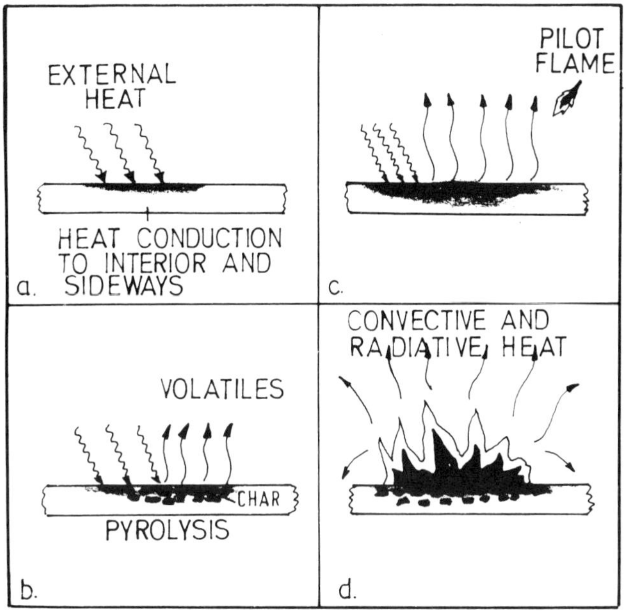

Fig. 4.2 — Ignition of solid material: (a) external heat reaching the surface, (b) pyrolysis begins, (c) ignition of flammable mixture, (d) spread of fire by thermal feedback.

presence of oxygen in air. Pyrolysis can be enhanced if the heating rate is high and the ignition source is strong.

The evolved gaseous products mix with air and form a flammable mixture. At temperatures around 580 K the mixture is easily ignited by a small pilot flame (Fig. 4.2(c)). If flame is absent, the mixture may spontaneously ignite, although at considerably higher temperatures, *ca*. 770–880 K. At this point the overall pyrolysis reaction becomes exothermic.

The diffusion flame thus formed rapidly covers the surface of the pyrolyzed material, heats it up further by thermal feedback, thus supporting its combustion (Fig. 4.2(d)). If the external ignition source is now removed, the combustion will continue unless the resulting heat loss is too great and reverses the reaction again to the endothermic range. Whether this will occur or not depends on a number of factors, of which the inherent physical characteristics (expressed as thermal absorptivity), the thickness of the material and its orientation are among the most important. Thus, materials which are easily ignited usually easily sustain the spread of flame. In fact the spread of fire can be regarded as a succession of repeated ignitions.

Thin materials cannot readily dissipate the incoming heat and thus are heated up quicker, while thick materials, especially if good conductors, conduct the heat from the surface to the interior, consequently reducing the surface heating and the rate of fire spread.

The infleunce of the orientation of the material in space is well known. Spread along vertical surfaces is fast, as convective heat transfer considerably extends the

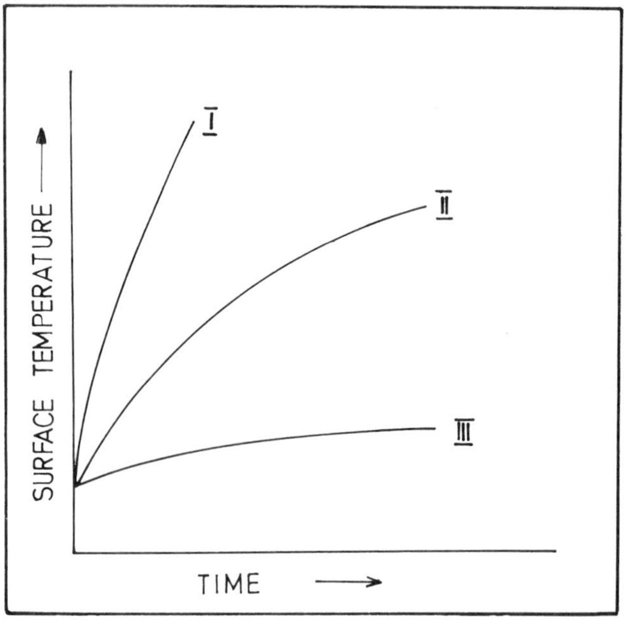

Fig. 4.3 — Surface temperature–time curves for materials of different thermal absorptivity $(K\rho C)^{1/2}$. Thermal absorptivity follows the order I<II<III.

heating zone and flames directly touch the surface. On the other hand, horizontal spread is usually slow, since the heat is really convected away. The adjacent surface is heated only by heat radiation from the flame or heated neighbouring materials or objects, or by heat conduction if the material is rather thick.

The direction of fire spread is also dependent on some external factors, such as the intensity and the duration of the external heat flux, as well as on environmental conditions (Rieber 1983).

The above qualitative discussion referred only to combustion as a form of burning accompanied by the appearance of flame. Other forms of burning, such as smoldering and glowing, appear in practice less frequently. Usually, they commence more slowly and go on much longer. Therefore, they are less threatening to human life, and will not be dealt with further. Because the terms 'burning', 'combustion', 'smoldering', etc., are still frequently misused, an unequivocal terminology has been called for (Harmathy 1984).

Once a fire has been ignited and begun to spread, its further progress, and its effects on people and property, will largely depend on whether it occurs in the open or inside an enclosure. A separate discussion of these two types of fires is therefore warranted.

4.3 FIRES IN THE OPEN

Knowledge of the dynamics of fires in the open is of practical interest for the prediction of possible material losses during a fire. However, the loss of human life is likely to be less in comparison with the possible loss caused by fires in enclosed spaces where people have a much greater chance of being trapped inside.

The process of fire growth can be divided into three distinct phases.

In the first phase, fire spreads and burning occurs on almost all the available fuel surface (at least 80%). The quantity of heat evolved in the unit of time depends on the heat of combustion of fuel, on the surface area of the fuel, on the mass rate of fuel burning per unit area and on the completeness of burning. The duration of the first phase is determined by the burning rate and the rate of flame spread and varies from 2% to 30% of the entire fire period.

In the second phase the mass rate of fuel burning and the evolved heat reach their maximum values and remain constant. Strong air currents are entrained into the turbulent flame and flying brands from the fire can be transported great distances, threatening nearby objects.

In the third phase the rate of burning diminishes and eventually stops when the oxidation of the solid residue has been completed.

Graphical representation of various phases of the course of a fire can be made by plotting a fire parameter (e.g. temperature, heat evolved, mass loss, etc.) against time. The plot of temperature vs. time is shown in Fig 4.4.

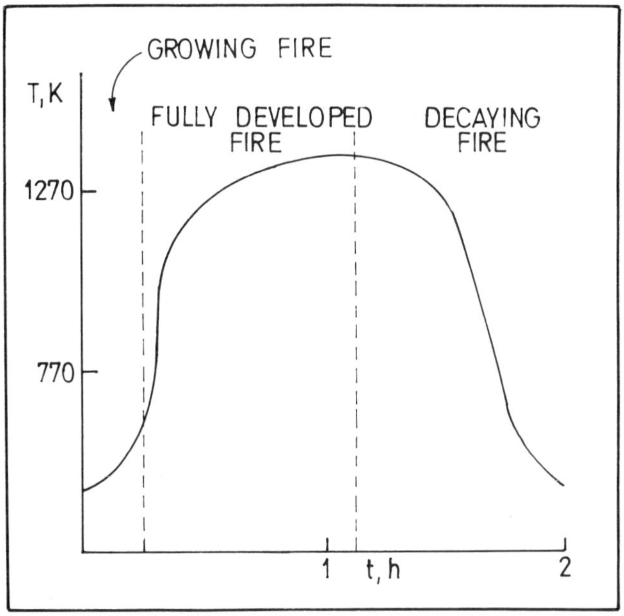

Fig. 4.4 — Theoretical temperature — time profiles of fire showing three distinct phases of the course of fire.

The average rate of burning may be expressed as (Kimstach *et al.* 1984):

$$R = G_i\, G_o/\tau_i A_F \tag{4.3}$$

where R is the rate of burning of the fuel from the given area and in the given time, (kg/(m^2×s)), G_i is the fraction of fuel mass burnt at the moment τ_i, G_o is the original mass of fuel (kg), τ_i is the duration of the fire, corresponding to the period of burning of the G_ith fraction of the original fuel mass (s), and A_F is the area under fire, in which burning of the fuel takes place (m^2).

Values of the average rate of burning of solid fuels range from 0.004 for peat to 0.0185 kg/m^2×s) for wood (with 8–10% moisture). Corresponding figures for liquid fuels are considerably larger (for example, diesel fuel burns at an average rate of 0.055 kg/m^2×s)).

If the average rate of burning, R, and the characteristic dimension of the flame base, d (m), are known, the flame height, H_f, in the absence of wind, can be calculated as:

$$H_f = C(Rd)^{2/3} \tag{4.4}$$

where C is a coefficient dependent on the fuel used (for wood C=16.4).

It is obvious that with an increasing rate of burning and with an increasing characteristic dimension of the flame base the flame will rise to greater heights. The characteristic dimension of the flame base in the case of rectangular fires is equal to the square root of the area under fire, while for round, or similarly shaped areas, it is equal to the diameter of a circle of the same area.

In the case of wind the flame elongates and becomes inclined. The flame height then depends on the angle of flame inclination, α (Fig. 4.5), and may be expressed by

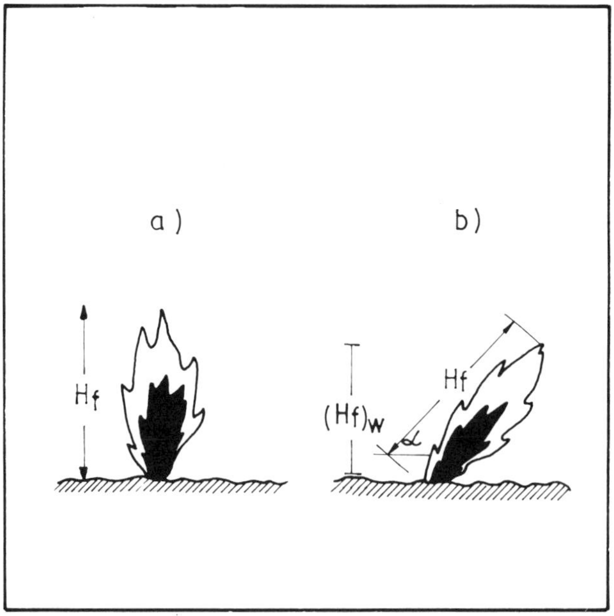

Fig. 4.5 — The height of a fire in the open: (a) in calm weather, (b) in windy weather.

equation (4.5):

$$(H_f)_w = H_f \sin \alpha \tag{4.5}$$

Here, H_f is practically the same as the value which would have been obtained were the burning taking place in the absence of wind, since the rate of burning is directly dependent on wind speed.

The height of the flames and the angle α are important variables for the computation of the distance necessary for the separation of the exposed material (structure) from the exposing open fire (see subsection 7.4.1).

The next important parameter that characterizes fires in the open is the intensity of heat radiation, Q, from fire to neighboring structures or materials. The quantity of evolved heat depends on the intensity of the radiation from the flame, I_o, on the absorption coefficient of the environment, β, on the distance of the flame from the exposed structure, r, on the coefficient of irradiation, ϕ, and on the area of the flame oriented to the exposed structure, A_{fl} (equation 4.6)):

$$Q = I_o \, e^{-\beta r} \phi \, A_{fl}/\pi r^2 \tag{4.6}$$

Values of β range from 0.2 to 0.3 for polluted air, and may rise to as high as 2.0 in case of the absorption of heat by fog. The intensity of the radiation from the flame, I_o, is governed by the well known Stefan–Boltzmann law by which I_o is proportional to the 4th power of the flame temperature [K]. The area of the flame may be approximately calculated as the product of the flame base and its height, corrected for the real appearance of the flame shape.

The average rate of spread of fire in the open, V_s, is a function of the duration of the fire, τ, and the total area under fire, A_F:

$$V_s = \left(\frac{1}{\tau}\right)^{-1} (A_F/\pi)^{1/2} \tag{4.7}$$

or

$$V_s = 0.564/\tau \, (A_F)^{1/2} \tag{4.8}$$

The distance to which fire in the open will spread depends mainly on the direction and speed of the surface wind and the degree of moisture in the burning material. If the wind speed is, V_w, then V_s can be calculated as:

$$V_s = 0.55 \, V_w^{0.7} \tag{4.9}$$

The spread of fire to the side perpendicular to the wind direction is slower and may be expressed as:

$$(V_s)_p = 0.45 \, V_w^{0.37} \tag{4.10}$$

Another approach to the problem of burning in the open has been presented by Harmathy (1972). His description refers to burning of piles of wood stacked in cribs. The rate of burning is a measure of the rate of volatile formation, which is in turn taken to be equal to the rate of decrease of the mass of the pile (an assumption not necessarily always correct). Experimental curves relating the change of mass of a burning pile as a function of time have shapes typical of thermogravimetric curves (Fig. 4.6).

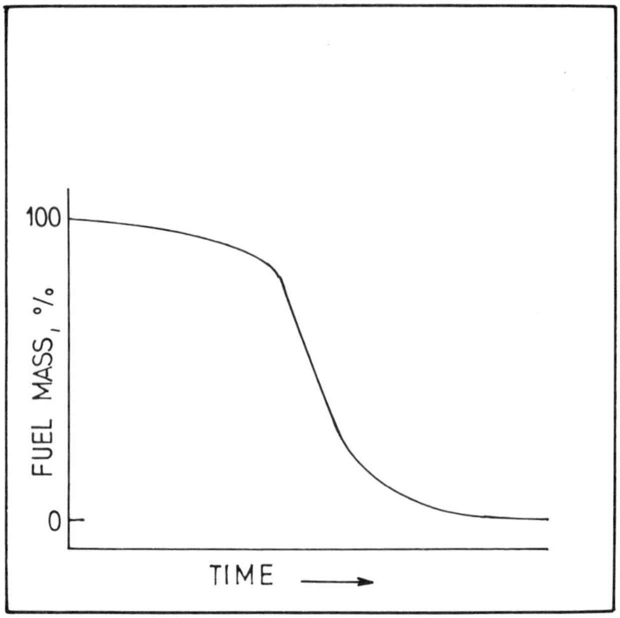

Fig. 4.6 — Mass change–time curve for a burning wood pile.

In such curves, three separate regions of mass loss can be observed. In the first region the mass of the pile reduces, at first rather slowly, then somewhat faster until about 20% of the original mass is lost. From that point on, burning progresses approximately constantly as a result of the constancy of the volatiles formation. When most of the volatiles have been driven off (at about 70% loss of the original mass of the pile) the rate of mass change slows down and eventually stops, leaving only partially oxidized char as residue.

Experimentally obtained mass loss curves have been mathematically modelled, using several simplifications (Harmathy 1972). Thus, the entire period of burning has been divided into two stages, designated as the stage of primary burning and the stage of secondary burning. The first stage covers the period of volatile formation, while the second stage refers to the final period of burning, characterized by the completion of char oxidation (it is assumed that during the first stage, half of the char has already been oxidized). The most important variables needed for the characterization of burning of wood piles in the open are: the rate of volatiles formation, U_v, the rate of mass loss R, the rate of heat evolution during the burning, Q, and the duration of burning τ. Their values can be computed using the following reasoning and expressions (Harmathy 1972).

The rate of volatiles formation, U_v, is directly proportional to the rate of mass loss, R. During the period of primary burning the following relation holds:

$$U_v = 0.932\ R \text{ kg/s} \tag{4.11}$$

The rate of mass loss due to char oxidation, R_o, is normally only a small fraction of the rate of mass loss during the period of primary burning, R, that is,

$$R_c = 0.068\ R \text{ kg/s} \tag{4.12}$$

The rate of heat evolution, Q, is dependent on the heat of combustion of both volatiles, ΔH_v, and char, ΔH_c, and on the rate of mass loss, R. For the duration of primary burning it can be expressed as:

$$Q = R\ (0.932\Delta H_v + 0.068\Delta H_c) \text{ J/s} \tag{4.13}$$

while during the period of secondary burning:

$$Q = R\ 0.068\Delta H_c \text{ J/s} \tag{4.14}$$

The duration of primary burning, τ, is directly proportional to the original mass of the pile, G_o, and inversely proportional to the rate of burning, R:

$$\tau = 0.936\ G_o/R \text{ s} \tag{4.15}$$

4.4 FIRES IN ENCLOSURES

Fires in enclosed spaces have been extensively studied, both experimentally and theoretically. The overwhelming majority of the research refers to the study of fires in residential compartments, as these enclosures are the most dangerous to people in the event of fire. Although considerable progress has been made, there are still many gaps in the knowledge of the dynamics of fires in enclosures which are filled with many 'ifs' and 'buts' and with empirically obtained facts.

From the aspect of life safety it is convenient to divide fires in enclosures into two periods. The first one is related to the ignition and spread of fire on individual objects in the enclosure, the so-called 'pre-flashover' period, and the other one refers to the rapid ignition and burning of all available combustible material ('flashover') and subsequent spread of fire outside the enclosure ('post-flashover' period). The first period is important for the safety of people within the enclosure, while in the second period, people in other parts of the building are also threatened, and the whole building may become structurally endangered.

4.4.1 Pre-flashover period

In the incipient phase of a fire in an enclosure its growth and spread are similar to what they would be in the open. However, the further development and course of the fire depend on a number of factors, among which the following are the most important: the total amount of combustible material within the compartment, including the lining material; the rate of heat radiation by the item first ignited, and the geometric and thermal characteristics of the compartment boundaries.

As will be shown, well furnished enclosures lined with combustible material spread fire more readily than sparsely furnished enclosures with incombustible walls and ceilings. Once an item in the compartment has caught fire, hot combustion products rise toward the ceiling, entrain surrounding air and migrate toward the ceiling and spread in the form of a buoyant plume. As hot gases and smoke accumulate beneath the ceiling a significant pressure difference gradually builds up and two distinct layers appear. The upper, hotter layer grows in size until it eventually reaches an outlet, usually the upper part of a window or door opening. As the burning of the item intensifies, the temperature of the hot layer rises, heating up the ceiling and walls. They, in turn, radiate heat energy back to the rest of the as yet unignited combustibles on the floor and in the lower part of the enclosure. Heating of the combustibles by radiation from the flames and by conduction through floors is of much lower intensity.

The intensity of the heat flux reaching the floor surface is dependent on the fire characteristics of the linings which cover the ceiling and walls and on the height of the ceiling. If the radiative heat flux reaching the floor level is Q, then its distribution along the floor of the enclosure may be qualitatively depicted as in Fig. 4.7.

Fig. 4.7 indicates that combustible ceilings and walls increase the heat flux, reaching floor level. This means that fire can spread more readily along floors if the ceiling and walls are lined with a combustible material. In meagerly furnished enclosures it may happen that the item on fire just burns itself out with no fire spread to other combustibles. On the other hand, in more heavily furnished enclosures, such as living rooms and bedrooms in a house, the chances are that fire will spread more easily and that the whole enclosure will catch fire. The as yet unignited combustibles on the floor and in the lower level of the enclosure heat up more and more by the radiative heat from the hot upper layer and ceiling. At some point, almost all of them heat up to their ignition temperatures, and on exceeding them, fire spreads very rapidly over all the enclosed space. This transition from a localized fire to full enclosure involvement is referred to as 'flashover'. The succession of events leading to flashover is shown in Fig. 4.8.

The definition of flashover is not as yet agreed on by the fire community. Some of the most frequently quoted definitions are shown in Table 4.2.

4.4.2 Post-flashover period

The post-flashover period of fire in an enclosed space is often termed a 'fully developed fire'. All the important fire parameters such as temperature, heat

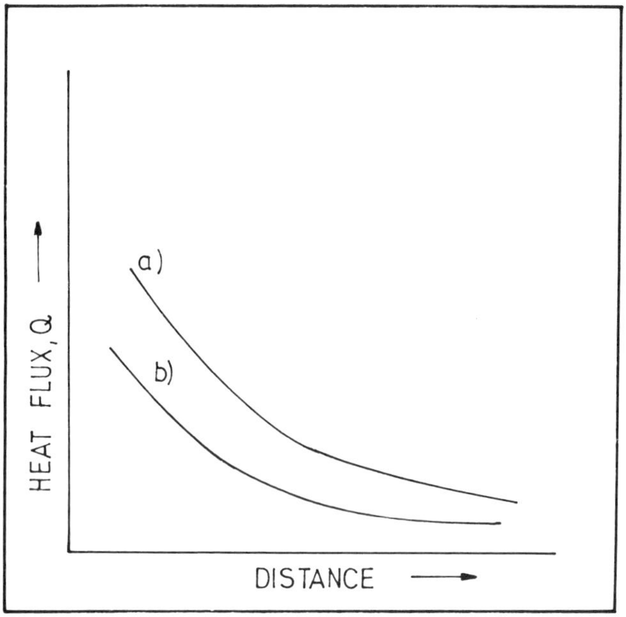

Fig. 4.7 — The distribution of radiative heat flux along the floor in an enclosure containing: (a) combustible linings, (b) noncombustible linings.

evolution, the rate of burning, smoke production, flame height, etc., are now at their maximum values.

It was thought for a long time, based on the results of research on fires in enclosures performed in the 1920s by Ingberg (Ingberg 1928), that the nature of fully developed fires is chiefly determined by the amount of combustibles present in the enclosure. In other words, the significance and role of only one reactant (fuel) in the basic fire reaction (see equation 4.1)) were recognized. Consequently, the concept of fire protection in enclosures by employing fire resistive boundary elements has been widely accepted and is used even today in many countries as a major component of fire protection.

Deeper understanding of the post-flashover period of fires has been advanced by a number of tests of fires in enclosures (Butcher *et al.* 1966, Gross & Robertson 1965, Kawagoe 1958) backed up by thorough theoretical studies of Harmathy (1972), Magnusson & Thelandersson (1970), Kawagoe (1967). Thomas *et al.* (1967) and others. A characteristic of these tests and studies is that the great majority of them referred to ordinary cellulosic materials only. It is not yet known whether the results obtained could be applied to other combustible solids, e.g. plastics, or some metals. Another important characteristic is the complexity of the experimental tests and studies arising from the complexity and inherently poor reproducibility of fire processes. Later, various researchers used various techniques and models in approaching and solving the problem.

Fig. 4.8 — Growth of fire in an enclosure leading to flashover.

In spite of these drawbacks some undisputed findings have emerged. One of the most important was that the nature of a fully developed fire depends as much on the second reactant in the fire equation, that is, the amount of air present and entering the enclosure, as on the amount of fuel involved in the fire. It has enabled important variables in fire such as the rate of burning, temperature within the enclosure, the rate of heat evolution, and the duration of the fire to be predicted fairly accurately. The discussion that follows is based in part on the excellent papers by Harmathy (1972, 1976, 1983), Lie (1974), and Thomas *et al.* (1967) and Drysdale's book (Drysdale 1985).

A typical enclosure studied is shown in Fig. 4.9.

If air enters the enclosure only through a broken window then the mass flow rate of air, U_a, at gas temperatures higher than 573 K is determined by a ventilation parameter θ, which includes the geometry of the opening (the height, h_w, and the area, A_w, of the effective window opening)

Table 4.2 — Definitions of flashover

Definition	Author
1.25 W/cm^2 radiation flux incident on the floor	Waterman (1972)
The moment when all exposed flammable substances contribute to the production of gaseous fuels	Thomas (1974
The point when flames emerge from the doorway and ceiling gas temperature is 600°C or more	Hagglund *et al.* (1974)
A critical state in the fire growth in a room which marks the difference between a fire that stays relatively small and confined to its initial surroundings and a fire that reaches its fully developed state	Quintiere (1977)
If the ratio of the rate of heat evolution by the fire and the product of thermal inertia of the compartment lining materials, total surface area of the compartment boundary and ventilation factor is greater than or equal to 8100	McCaffrey *et al.* (1980)

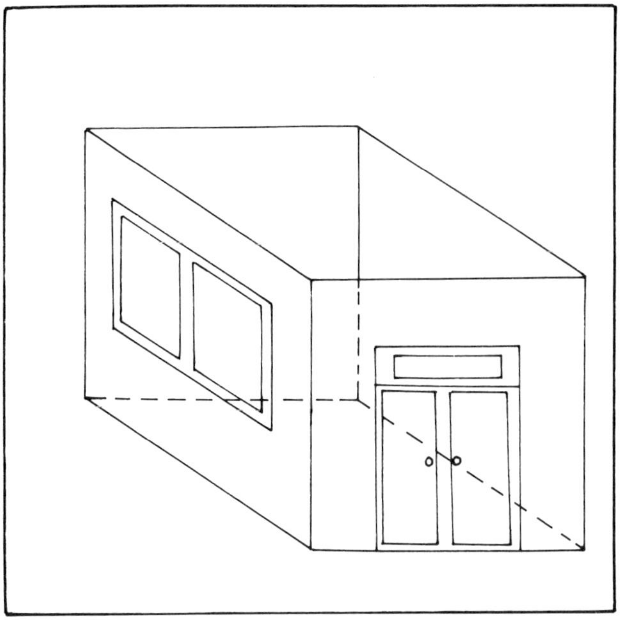

Fig. 4.9 — A typical fire enclosure.

$$U_a = 0.145\ \theta \text{ kg/s} \tag{4.16}$$

where θ is defined as

$$\theta = \rho_a\, g^{1/2} A_w\, h_w^{1/2} \tag{4.17}$$

As to the combustible material, what is important is not only its quantity, expressed as the total fire load, G_o, prior to fire, but also the extent of the availability of its surface area to air, the so-called 'free' surface, A_f. The total fire load is equal to the product of the specific fire load F and the floor area of the compartment, A_F, that is, $G_o = F A_F$. The free surface area is normally just a fraction, ϑ, of the total fire load ($A_f = \phi G_o$). In ordinarily furnished compartments ϑ is, on the average, equal to 0.25 m^2/kg. If the total fire load is expressed as the product of the specific fire load and the floor area of the compartment then:

$$A_f = \vartheta\, F\, A_F \text{ m}^2 \tag{4.18}$$

Now, when both of the reactants in the fire reaction are somewhat better characterized, it is possible to speculate on the nature of the fully developed fire. According to the previously mentioned Le Chatelier's principle, when one of the reactants is in excess then the concentration of the other one controls the reaction rate. At some optimum concentration ratio, the rate of the reaction is equally dependent on either of the reactants. This optimum ratio of air and fuel, θ/A_f, is termed 'critical', and for a compartment fire it has a value of about 0.26 kg/m^2 s^{-1}. The corresponding 'critical' air flow is then:

$$(U_a)_{cr} = 0.009\ G_o \text{ kg/s} \tag{4.19}$$

When the flow of air entering the compartment is lower than $(U_a)_{cr}$, the rate of burning, R, is determined mostly by the value of U_a, and can be approximated as:

$$R = 0.163\ U_a \text{ kg/s} \tag{4.20}$$

Such a fire is said to be 'ventilation controlled'. Ventilation-controlled fires are characteristic for compartments with relatively small openings. However, if the compartment is heavily loaded with the combustible material (in the range 40–100 kg/m^2) then fire in such a compartment will also be ventilation controlled, regardless of the size of the openings. In extreme cases, the air flow may not be enough to sustain a flammable burning, and the fire smolders, or dies out.

On the other hand, when the rate of air flow is larger than $(U_a)_{cr}$ then the free surface area of the combustible materials becomes the controlling factor of the rate of burning. Such fires are said to be fuel surface controlled. The rate of burning of such fires may be roughly estimated as:

$$R = 0.0062\ A_f\ \text{kg/s} \tag{4.21}$$

The duration of a fire, τ, is directly proportional to the amount of fuel involved, G_o, and inversely proportional to the rate of burning, R:

$$\tau = 0.932\ G/R \tag{4.22}$$

Since these two variables strongly depend on the ventilation factor, U_a, the duration of the fire will also depend on it. Thus, for the ventilation-controlled regime, it can be expressed as $\tau = 5.72\ G_o/U_a$, while for the fuel-surface controlled regime it can be shown that τ is practically independent of the fire load and is approximately equal to 0.3 h. However, it should again be stressed that the above conclusions refer to typically furnished residential and office occupancies only. The situation may be, and often is, quite different in the case of fires in large industrial structures characterized by sometimes enormous fire loads and by specific geometries of enclosures.

The next important parameter of the fully developed fire is the average fire temperature, since the temperature course of a fire in an enclosure forms a basis for the fire-resistive design of buildings. One would expect that the average temperature is influenced by the specific fire load and the rate of air flow in such a way that the larger these two variables are, the higher the average fire temperature. In reality, curves such as those shown in Fig. 4.10 have been theoretically calculated and experimentally confirmed.

They show that the average fire temperature does indeed increase with increasing ventilation and fire load, but only up to a point beyond which the temperature decreases as the ventilation increases. The rationale of this fact is that at high ventilation the burning rate and flame height become so large that fire has to spread outside, to open space or to other neighbouring compartments. Consequently the fire products release a significant part of their chemical energy out of the compartment, thereby lowering its average temperature. It follows that maximum fire temperatures mostly occur at relatively low ventilation, which means that ventilation-controlled fires are the most destructive. On the other hand, well ventilated fires, that is, fuel-surface controlled fires, are relatively cooler, and are of rather short duration. This sharply contradicts the popular misconception that fires in enclosures are either short and hot, or long and of lower temperatures.

Since the fire load in an enclosure in a building cannot be predicted because it often changes with time and place, the temperature course of the corresponding fire cannot be known in advance. Among a large, almost infinite, number of possible temperature courses of practical interest there are those that reflect the worst possible fire scenario (poorly ventilated fires involving high fire load) which would

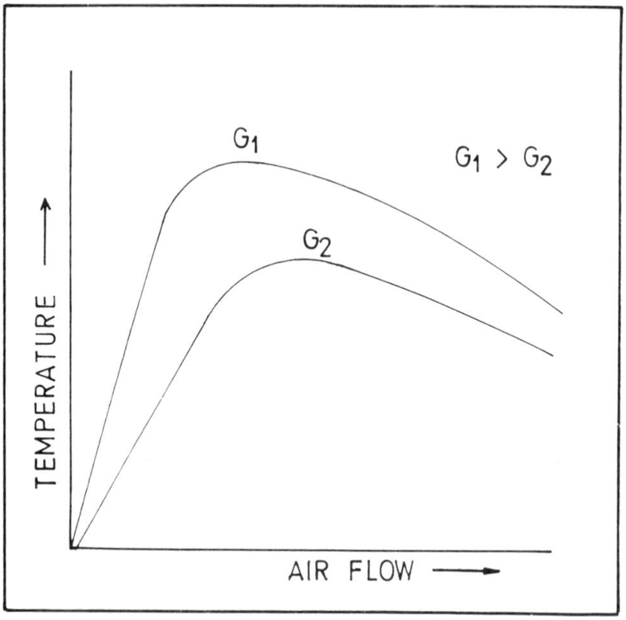

Fig. 4.10 — Fire temperature–air flow curves for enclosures containing two different fire loads ($G_1 > G_2$).

not be realized during the lifetime of the building. Analytical expressions which define these curves have been derived (Lie 1974).

The recognition that the nature of fire in an enclosure is dependent not only on fire load but also on ventilation has led to a different explanation of the mechanism of fire spread. The customarily accepted fire protection philosophy has been that fire spreads from the enclosure of origin to a neighbouring room by heat conduction through the boundaries or by structural failure of a boundary element. However, using the above reasoning it could be easily proven that were fire spread to occur only by these two mechanisms, a great majority of fires in ordinary enclosures would last for less than one hour. In such a short time no major structural failure of the ordinary building elements and construction is likely. The fire spread most often observed in reality is a consequence of heat convection and radiation from advancing flames coming out through enclosure openings, such as doors, windows, vertical shafts, and openings in ceilings and directly touching or irradiating a combustible material. These findings have a potentially far-reaching impact on conventional architectural fire protection design (see section 7.1).

4.5 COMPUTER MODELING OF FIRES IN ENCLOSURES

The most logical way of studying fire in an enclosure and its spread beyond enclosure boundaries would be by performing fire experiments. Indeed, a large number of experiments have been performed on both full- and small-scale enclosures and/or samples of materials representative of furnishings of some typical enclosures (for example, residential and office compartments). In this way a great deal of useful information on fire and its nature in enclosures has been collected.

However, experimental results obtained usually refer to one particular fire scenario only, with a predetermined arrangement of combustible materials, the position of ignition source, thermal properties of internal linings, the size of the openings, etc. It would be prohibitively expensive to carry out additional full-scale fire experiments on the same enclosure differing only in minor details. Yet, even such minor changes of detail can lead to different experimental results and conclusions.

On the basis of numerous experiments on fire in enclosures, a theory of fire ignition, growth and spread has emerged. Although it is still in its early stages because of the enormous complexity of phenomena taking place in a fire it has progressed to the point it can predict fairly well the growth and spread of fire in an enclosure. The considerations in sections 4.3 and 4.4 have been founded mostly on the basic principles of this theory.

The predominant form of fire theory has been the mathematical modeling of the processes that govern fire dynamics. Although fire modeling has been steadily growing as an expanding research area, the number of researchers and centers where it is being performed is still rather small. Major research activity in this field is taking place in the United States, Japan, the UK, and some other European countries.

A considerable number of models have been developed for different purposes. They can be grouped into two classes, deterministic and stochastic (Thomas 1983). Deterministic models use basic postulates of fundamental sciences to predict a particular fire parameter, for example smoke concentration in the initial phase of fire growth, while stochastic models regard fire as a randomly determined sequence of processes which are best described statistically, that is, assuming the probability of a particular fire parameter to have a certain value. Both classes include a variety of implicit assumptions and mathematical techniques. The ensuing computational complexity makes the use of a computer a necessity. The majority of fire modeling has been done using deterministic models.

Ideally, a fire model should be able to predict the development of a fire in an enclosure and its spread in the building so that appropriate preventive measures for life and property safety could be applied by designers and regulators. In reality, this goal has not yet been achieved, at least not completely, although the advances in this field are impressive. The usual practice has been to develop models for separate periods of fire growth, for example pre-flashover, fully developed and decaying periods, depending on whether the primary interest is in saving lives or in averting significant property losses. A detailed discussion of fire modeling lies outside the scope of this book. Therefore, only some of the most important works will be briefly mentioned.

Quintiere (1977) has reviewed research on fire growth and spread in enclosures and has developed a quasi-steady mathematical model of fire growth in an enclosure. The enclosure is divided into two distinct layers, or homogeneous zones, separated by a sharp thermal discontinuity. The upper zone contains hot gaseous products and flames and the lower zone is filled with cold air drawn in through the opening. The model employs characterization of the physical processes of fire growth and spread by mathematical equations. However, some important chemical aspects of fire growth, such as stoichiometry of fuel combustion and the combustion outside the enclosure, are left unresolved because of their complex nature. The principal variables that influence the results are the thermal properties of combustible materials, the location and size of the fire origin, and the geometry of the enclosure. Even though in reality the zones are neither homogeneous in temperature nor homogeneous in composition (Quintiere 1981) the assumption of two layers is useful, as it simplifies the model and enables fairly correct predictions of fire growth in an enclosure to be obtained (Hagglund 1984).

Emmons (1978) has produced a computer program for the prediction of fires in buildings. The enclosure in which the fire has originated and the remainder of the building are divided into zones which are described as 'inputs', 'outputs' and 'storage' of the process variables. The zones are connected by equations expressing the mass and heat balances. The comparison of experimentally obtained and computed data has been satisfactory, revealing the good predictive power of the program. More advanced versions of this program have been developed. Thus, analysis of fire behavior in an enclosure is best performed employing the program Mark V (Butcher & Parnell 1985), while fire growth and spread in a multi-enclosure structure can be simulated using the NBS/Hardwarr Mark VI program (Rockett & Morita 1985).

Harmathy & Mehaffey (1983) thoroughly reviewed the existing mathematical models of post-flashover fires in enclosures which are of principal interest primarily for those concerned with the structural safety of structures. Each of fourteen models was examined and classified with respect to important aspects such as the utility in design, interpretation of fire duration, the way process variables are treated, characterization of the destructive potential of fire, structure of the system, the rate of burning, type of fuel, with the emphasis on cellulosic fuel, constancy of the rate of burning of cellulosic fuels, stoichiometry of combustion, the possibility of combustion of volatile pyrolysis products outside the enclosure, the coefficient of heat transfer to the enclosure boundaries, the thickness of enclosure boundaries and thermal properties of the enclosure boundaries.

One important piece of information that is expected from a model is the potential of an enclosure fire to spread outside the boundaries. Harmathy and Mehaffey noted that almost all of the models examined regard the destruction of enclosure boundaries and the conduction of heat through the boundaries as the principal mechanisms of fire spread. However, fire practice definitely proves that the most frequent means of fire spread is by convection through enclosure openings, and only rarely by destruction of the boundaries. A model has been developed (Mehaffey & Harmathy 1981) which is capable of predicting the potential of fires to spread by both convection and destruction, and which introduces a theorem of uniformity of normalized heat load on the enclosure boundaries. On the basis of this theorem, quantitative expressions for the potential of fire spread have been developed. By employing an improved version of the model developed at the National Research Council of Canada (Harmathy 1986), the growth of fire on account of convective flow of hot fire gases and the evaluation of the fire resistance of enclosure boundaries during the burning of liquids and thermoplastic solids can be made.

The achievements and shortcomings of computer modeling of fire growth have been reviewed (Parikh & Beyreis 1985, Thomas 1983, 1985, Emmons 1981).

4.6 DYNAMICS OF EXPLOSIONS

The subject of explosion fundamentals and explosion hazards of materials is covered in more detail in Chapter 2. Therefore, only those parameters of explosions responsible for destructive effects will be briefly discussed here.

On first sight the term 'explosion dynamics' might seem misleading, as explosions, by definition, imply very fast, in most cases chemical, transformations accompanied by a sudden release of energy in the form of pressurized gas. However, for the effective detection and suppression of an explosion, as well as for an appraisal of explosion damage, it is important to have information on the burning velocity and the flame speed, the maximum pressures developed, the rate of pressure rise, and the duration of the excess pressure. These parameters also determine the dynamics of an explosion.

Burning velocity

In ordinary combustion of gases and vapors, flames propagate through the reaction mixture, with velocities around 0.5 m/s (Marshall 1987). These velocities are too slow to cause an appreciable pressure rise in unconfined systems. Of course, if the reaction is confined, gaseous products of reaction accumulate and at some point the resulting overpressure exceeds the design strength of the enclosure, causing its rupture (Harris 1983).

The combustion velocity of gases mixed with air depends on the mixture composition, the initial temperature and pressure, and the chemical purity of the mixture. The fastest combusation occurs at concentrations around the stoichiometric ratio of gas and air, which corresponds to the middle point of the flammability range (Cruice 1986). Below the lower and above the upper flammability limits the mixture will not burn, a fact which points to one of the methods of explosion protection (see section 3.1).

The effects of initial pressure and temperature of the flammable mixture on burning velocity are fairly straightforward. At higher pressures and temperatures, molecules of the reactants are closer together and the reaction rate is therefore faster. The flammability range of the mixture widens and the overall explosion potential increases. Generally, the higher the initial pressure and temperature of the mixture, the higher the combustion velocity (Fig. 4.11).

In gas explosions the burnt gas pushes the flame front along. The velocity of the flame front, S_f, is influenced by the geometry of the enclosure, stoichiometry of the reaction and the ratio of the flame temperature, T_f, and the initial temperature, I_i (Harris 1983). The latter two, together with the mole ratio of reaction products and reactants, N_m, define the 'expansion factor', $E\,[E = (T_f/T_i)/N_m]$, which has a value of around 8 for confined and 2 for unconfined reactions, respectively. Consequently, the flame speed, S_f, in explosions exceeds the normal flame speed, S_o, by the same factors (Marshall 1987).

The burning velocity of the mixture also depends on the chemical purity of the

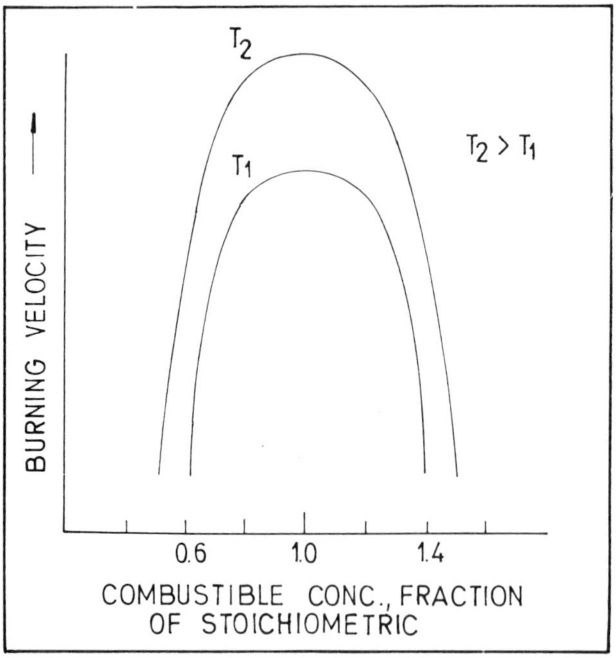

Fig. 4.11 — Dependence of the combustion velocity on mixture composition at two different initial temperatures of the mixture.

mixture. The effects of impurities, such as traces of moisture, dust, soot, etc., are unpredictable. Their presence may enhance or inhibit the reaction rate.

The reaction front in the detonation of dense (solid or liquid) explosives spreads through the reaction mass at roughly the velocity of sound in the material, which is of the order of several km/s. In deflagrations of a flammable material dispersed in air the corresponding velocities at maximum approach the speed of sound in air, or 330 m/s (Baker *et al.* 1983).

Maximum pressure

Since explosion of a combustible material liberates gaseous products, the pressure developed can be, in principle, readily calculated using the well known gas laws, provided that the initial temperature, pressure and volume are known. Knowledge of maximum pressure is important when it is intended to contain the explosion within the enclosure as a method of protection.

It can be theoretically shown that under adiabatic conditions the maximum pressure that can be developed in a confined deflagration is approximately 800 kPa (Marshall 1987). In practice, lower values of P_{max} are obtained owing not only to the deviation of real gases from their ideal state, but also to the influence of a number of variables, such as the chemical structure of the material, the composition of the mixture, the particle size of the material (if a flammable dust or mist), the volume of the enclosure, the degree of the turbulence of the mixture, the presence of moisture and inert substances, and the nature and location of the ignition source (Peterson 1986). The dependence of P_{max} on some of these variables is shown in Fig. 4.12.

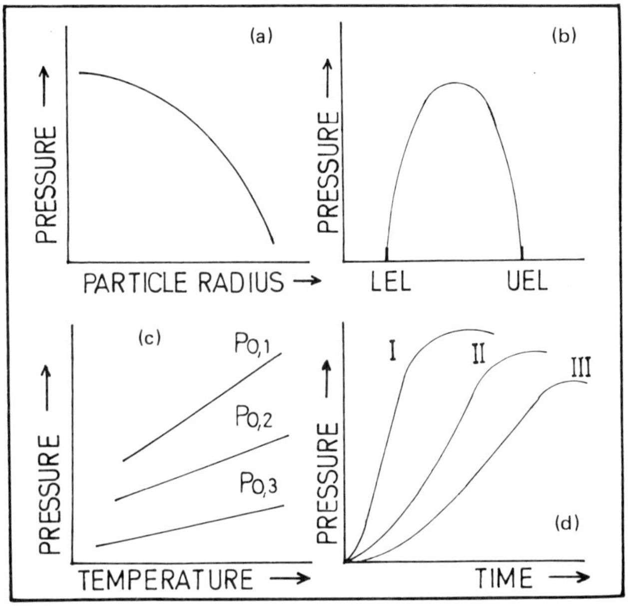

Fig. 4.12 — Dependence of P_{max} on (a) particle radius of dust, (b) explosive range (LEL and UEL — lower and upper explosive limits, respectively, (c) initial pressure ($P_{o,1} < P_{o,2} < P_{o,3}$), and (d) intensity of the ignition source (I > II > III).

Fortunately, only rarely do all these variables simultaneously attain values which are optimal for the development of the maximum pressure.

On the other hand, detonations, especially in condensed phase, produce much higher pressures, which may reach, under certain conditions, values as high as 20–40 GPa (Baker *et al.* 1983). Such high pressures exclude conventional venting methods as a method of protection and are responsible for the well known destructive effects of detonations.

Rate of pressure rise

The rate of pressure rise is one of the key explosion parameters. It is used mostly for the design of explosion-relief venting of industrial plants (Palmer 1973). Enclosures that are likely to experience deflagrations with a high rate of formation of combustion products, that is, high rates of pressure rise, need to be vented using larger venting areas. In cases where the rate of pressure rise is extremely rapid, explosion venting may be impractical, demanding alternative explosion suppression methods.

Whereas the maximum explosion pressure essentially is not dependent on the enclosure volume, the maximum rate of pressure rise, defined as dP/dt in Fig. 4.13, is.

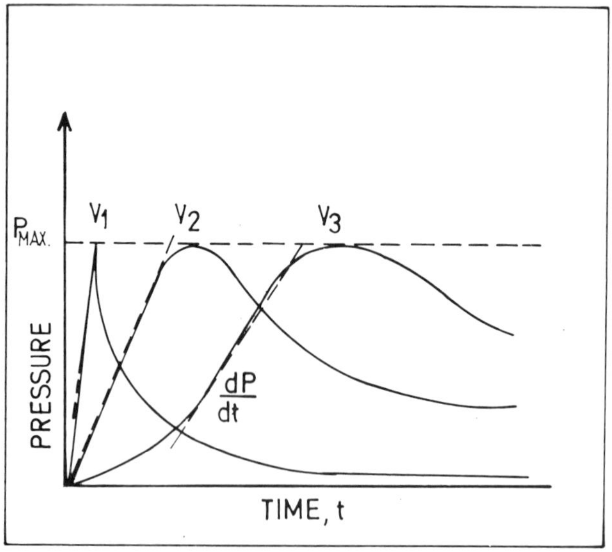

Fig. 4.13 — P_{max}–time curves for enclosures of different volumes ($V_1 < V_2 < V_3$). Rates of pressure rise, dP/dt are shown as slopes of the corresponding curves.

The quantitative relationship between the maximum pressure rise and the volume of the enclosure obeys a cube law (Bartnecht 1981).

$$(\mathrm{d}P/\mathrm{d}t)_{max}\, V^{1/3} = K \tag{4.23}$$

Values of K are used as a basis for the assessment of explosibility and classification of combustible dusts (Marshall 1987).

Experimentally obtained values of the maximum rate of pressure rise for industrial powders and dusts are in the range 100–160 000 kPa (Field 1983).

Duration of excess pressure

The destructiveness of an explosion is associated not only with maximum pressure and the rate of pressure rise but also with the length of time the excess pressure is exerted on the walls of the enclosure. The total impulse of explosions may be readily calculated as the area under the time–pressure curve. For practical purposes only, values of the positive impulse are important (Baker 1973), since underpressure, which defines the negative phase, is normally very small, always less than 100 kPa. The positive phase duration of explosions is expressed in milliseconds. Explosions having higher average rates of pressure rise may actually be less destructive than explosions which generate lower overpressures, since the duration of the latter is longer and the total impulse exerted is higher.

LIST OF SYMBOLS

A_F — area under fire (floor), m^2
A_f — free surface of fuel, m^2
A_{fl} — flame area, m^2
A_w — opening (window) area, m^2
d — characteristic dimension of flame base, m
F — specific fire load, kg/m^2
G_o — original mass of fuel (total fire load), kg
G_i — fraction of fuel mass burnt in time
g — gravitational acceleration, 9.8 m/s^2
H_c — heat of combustion of char, J/kg
H_f — flame height, m
$(H_f)_w$ — flame height in wind, m
H_v — heat of combustion of volatiles, J/kg
h_w — height of the opening (window), m
I_o — radiation intensity from the flame
Q — rate of heat evolution, J/s
R — average rate of burning, or rate of mass loss, kg/s
R_c — rate of char oxidation, kg/s
r — distance from the flame to exposed structure, m
U_a — rate of air flow, kg/s
$(U_a)_{cr}$ — critical air flow, kg/s
U_v — rate of volatile formation (equal to R), kg/s
V_s — average spread of fire in the open, m/s
(V_s) — spread of fire perpendicular to wind direction, m/s
V_w — wind speed, m/s

5

Fire and explosion detection and alarm systems

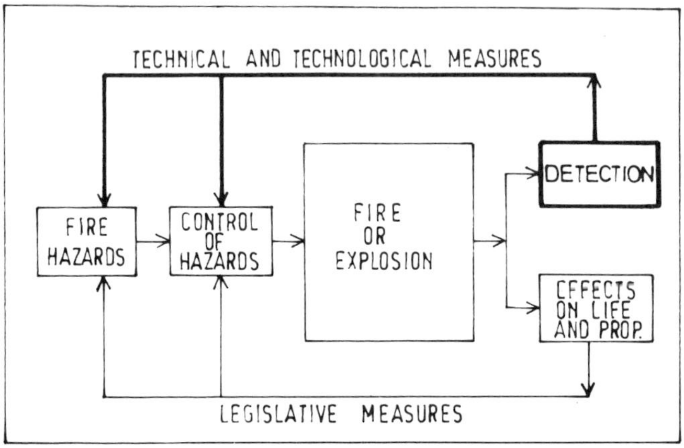

In the system approach to fire and explosion protection the component of the system dealing with the detection and alarm of fires and explosions and subsequent response action is of the utmost importance. The position of this component in the system, and its primary relationship with the blocks describing the dynamics of fires and explosions, the effects of these events on people and property, and the sources and control of fire and explosion hazards, clearly demonstate that if the protection of life and the prevention of property damage are to be secured, fires and explosions have to be detected and acted on at the earliest stage of their initiation and growth, when the fire is small in size and can be relatively easily extinguished. This statement has been repeatedly proven in practice. Properly designed, installed, operated and maintained, fire and explosion detection and alarm systems can substantially reduce fire losses.

For instance, in an analysis of the effectiveness of fire detectors and automatic sprinklers on human safety in the USA, Budnick (1984) has shown that the rate of fire deaths and injuries is considerably smaller in residential occupancies furnished with fire detectors and automatic spinklers. Since smoke detectors have become mandatory in residential occupancies in the USA, the annual rate of fire fatalities has declined from a maximum of *ca.* 10 000 to *ca.* 6000 (Walsh 1986).

Ramachandran & Chandler (1984), evaluating economic values of early detection of fires in industry, have calculated that average savings per fire due to early fire detection in the timber and furniture industry in the UK, in comparison with fires not discovered at ignition, ranged from 47% to 53%, depending on whether detectors were or were not directly connected to the fire brigade. A similar analysis of fires in the textile industry (Peissard 1983) revealed that around 40% of all fires in that industry were discovered at an early stage with the help of automatic fire detectors. The early detection and prompt extinguishment in such cases had reduced the size of the burnt area to an average of 17 m^2. When fires were detected at a later stage the size of the burnt area averaged 52 m^2, that is, *ca.* three times larger. Corresponding Swiss data on fires in the textile industry demonstrated a minimal material loss of under 1000 Swiss francs in almost 90% of all fires in premises protected by automatic fire detectors (Peissard 1983).

The above examples unquestionably substantiate the benefits of automatic fire detection and alarm systems. Therefore, it should be clear that the expenditures needed for installing such systems in industrial and residential premises are really quite minor in comparison with the benefits they provide.

5.1 AUTOMATIC FIRE DETECTORS

The best and most universal fire detector is man himself. Human beings possess senses which can detect all principal combustion products. Flames and smoke can be seen by the eyes, heat can be felt by touch, and gases can be smelled through the nose. It has been known for a long time that fewer large fires occur in the presence of people who are on the alert. However, there are many situations and places where people are not, or cannot always be, present. In such cases it is necessary to use automatic fire detectors. They should be able to respond to an indication of fire, to generate a signal and to send it to an alarm system for corrective action.

There are many types of automatic fire detectors. Their principal features are shown in Table 5.1 (EN 54 1976). The classification of fire detectors according to fire phenomena is shown in Fig. 5.1, while classifications, based on other features, are self-evident from Table 5.1.

5.1.1 Heat detectors

Operational principle

The operation of heat detectors is based on a change in the physical or electrical property of a material or gas under the influence of heat which then initiates an alarm

Table 5.1 — Classification of fire detectors (EN 54 1976)

Feature	Detector classification
Fire phenomena	Heat, smoke, gas and flame
Response mode	Static, differential, rate-of-rise
Resettability	Resettable (self-, remotely, locally, nonresettable (with and without exchangeable elements)
Configuration	Point, multi-point, line
Detachability	Detachable, nondetachable

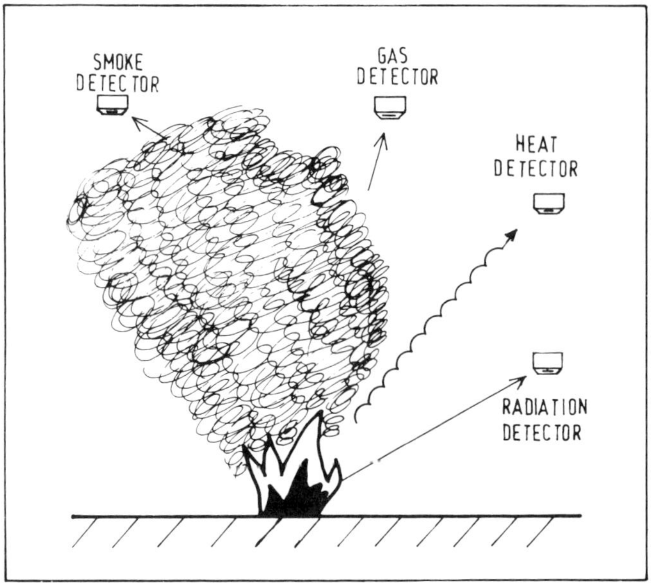

Fig. 5.1 — Classification of detectors according to fire phenomena.

(Zimmerman 1986a). Physical changes comprise thermal expansion, melting of a fusible alloy, softening of heat-sensitive insulation or volume expansion of a liquid, while the most often used electrical change is the change in electrical resistance (NFPA 72E 1984). These changes can be measured either at a fixed (maximum) temperature of the sensor, or in the surrounding air, or continuously. In the last case the detectors are activated when the rate of temperature rise exceeds a predetermined value (typically 7 K to 8 K per minute).

Application

Heat detectors respond to the convective heat liberated by the fire. Since heat intensity rapidly decreases with the distance from the fire center, heat detectors are best suited for fire detection in a small confined space, typically 15–30 m^2, with low ceilings, where at the outbreak of fire a fast rise in temperature is anticipated. They can also be used in the open, provided they are located directly over protected equipment, machinery, etc., where flaming fires could be expected. In principle, they could also be used for the protection of larger enclosures by employing a denser spacing, but this is not frequently practiced. Heat detectores are used in areas where relatively harsh ambient conditions, such as smoke, and humidity, preclude the use of other types of detectors, or where speed of detection is not of primary importance. They are therefore especially suitable for connection to automatic fire extinguishing systems (e.g. sprinklers). Since they can have a self-protecting circuit, heat detectors can be built in an explosion-proof design enabling their use in explosion-endangered atmospheres. Line-type detectors can be used for the protection of cable and traffic tunnels, conveyor belt systems, on airports and hangers, or in stacks, etc. False alarms with heat detectors are rare and appear, if ever, only when the detector is improperly located, for example near a heat source, such as heating installations, or in direct sunshine.

Despite a number of advantages (for example, the least expensive of all fire detectors, they have the lowest false alarm rate, they require either none or a very small power source, and have superior mechanical robustness) they gradually lose ground to the more versatile smoke detector (*Int. Fire Chief* 1986).

5.1.2 Smoke detectors

Operational principles

Smoke is an almost unavoidable product of combustion, since only rarely does a combustion reaction proceed completely stoichiometrically. The particulate nature of smoke permits its detection by measuring the interference effects of smoke particles on either light propagations (photoelectric smoke detectors), or through ionization current flow (ionization smoke detectors) (Zimmermann 1986a, NFPA 72E 1984).

Photoelectric smoke detectors can be of two types, depending on whether the obscuration or the scattering of the light beam is measured (Fig. 5.2). In the former case (Fig. 5.2(a)) the intensity of the light reaching the receiver is measured. In the absence of smoke, the light arriving at the receiver generates a steady photocurrent. When smoke is present it obscures the light and attenuates the current at the receiver. When the photocurrent falls below a predetermined value an alarm signal is actuated. The response sensitivity of the projected beam detectors can be adjusted in several defined stages. Signal strength reduction due to causes other than smoke, such as dust accumulation, detector aging or extreme environmental temperature changes, can be automatically compensated for. If the light beam is interrupted, for some physical reason, or the housing cover is accidentally opened or destroyed, or the limits of compensation are reached, a fault instead of an alarm signal is actuated. The signal intensity may also depend to some extent on the color of the smoke.

The photosensitive receiver can be placed at right angles to the direction of the light beam. Such a detector then measures the light scattered by the smoke particles

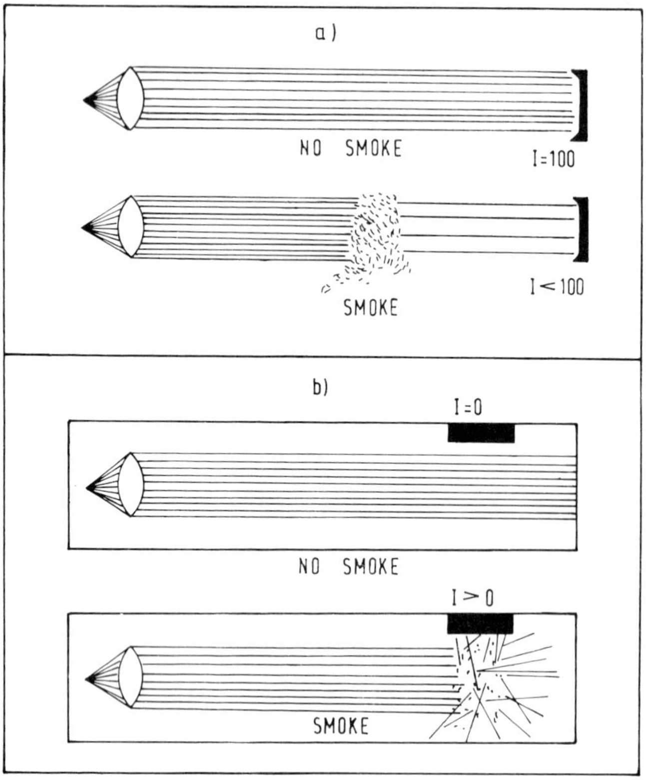

Fig. 5.2 — Response principle of photoelectric smoke detectors: (a) light obscuration and (b) light scattering.

(Fig. 5.2(b)) by analogy with the nephelometric principle of turbidity measurements in analytical chemistry. Interference from dust in air can be dealt with by adjusting the smoke entry, or by automatic adjustment of the signal threshold. Both light obstruction and light scattering smoke detectors are inactive if the radius of the smoke particles is less than the wavelength of the incident light, say less than 100 nm (Burry 1985). This is one of the reaons why ionization smoke detectors are more universally used.

The ionization smoke detector consists of a sensing chamber which contains a small amount of radioactive compound and two charged electrodes (Fig. 5.3). It is usually of the spot type (that is, mounted at a particular location). The radiation from the radioactive element keeps the air in the chamber permanently ionized, thus permitting a small steady current flow between the electrodes. If smoke particles enter the chamber they are electrically attracted by and become attached to the ions, thus reducing the conductance. When the conductance drops below a predetermined level, the detector responds. The ionizing current is dependent both on the radius and on the concentration of smoke particles entering the chamber (Helsper *et al.*

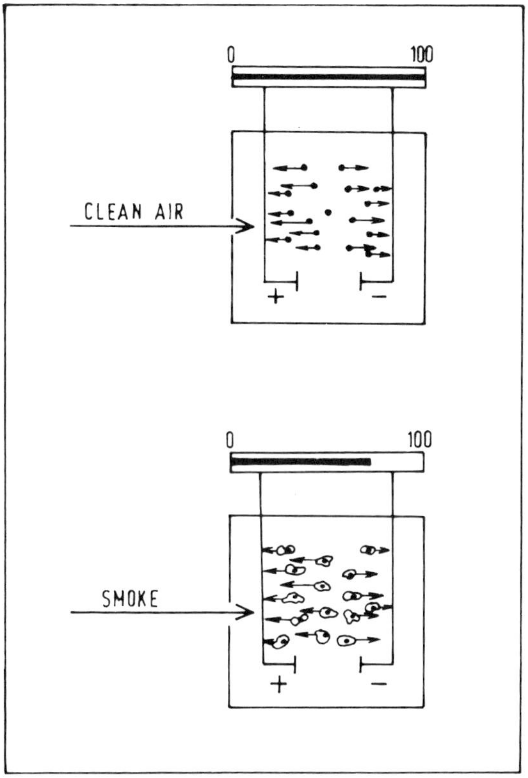

Fig. 5.3 — Response principle of an ionization smoke detector.

1983). The smaller the particles, the faster is the response. Ionization smoke detectors, as a class, operate considerably faster than photoelectric detectors. In experimental work involving imitations of smoky fires and using three types of fire detectors (ionization, photoelectric and heat detectors), the response times of the detectors ranged from 66 s (ionization detector), through 166 (photoelectic) to 240 s (heat detector) (Miyama & Saito 1982). Unfortunately, the ionization detector does not differentiate between particles of different origin. Any particle of the right size will actuate a signal, which means that ubiquitous particles such as dust, process fumes, etc., will trigger an alarm. Attempts have been made to obviate the problem by using two ionized sampling chambers, an outer and an inner one. While smoke particles can freely enter the outer chamber, the inner one is almost wholly inpenetrable. In the presence of smoke particles the electrical balance between the two chambers is distributed. The built-in electronic circuit responds to this change and produces an alarm.

The radioactive source which ionizes the air in the detector chamber should be a weak one, usually less than 3.7×10^4 Bq (1 μCi). Among a number of possible options, ^{241}Am, in the form of its oxide, AmO_2, has found widespread use. It has a half-life of *ca.* 240 years and emits α radiation which penetrates air only up to about 6 cm and thus can easily be shielded by a thin sheet of material such as paper. Thus the encasing of the detector serves as an effective barrier to radioactivity. At a distance of 0.5 m from an ionization smoke detector the level of radioactivity is so small that it is equal to the natural background (Lengauer 1986). Even in the worst case — breakage of the ionization detector in a fire — the risk of a fireman's direct contact with the radioactive sensor is very small.

Application

Smoke detectors are applicable to practially all kinds of fires and have become the predominant type of fire detector in both industrial and residential settings. In comparison with heat detectors, smoke detectors can protect considerably larger spaces because smoke does not dissipate as fast as heat in the same volume space. If flaming fires are expected, then ionization smoke detectors are somewhat better than photoelectric smoke detectors. The latter are best used for the detection of smoky fires (smoldering or low temperature fires) which generate relatively large particles. Spot-type smoke detectors are employed for the protection of relatively small enclosures, 20–50 m^2, while line-type detectors (which provide continuous detection along a path) may be used in enclosures with high ceilings, such as hangers, halls, tunnels, museums, corridors, etc.

The installation of smoke detectors should be subjected to careful fire engineering analysis of each case under consideration (Zimmermann 1986b, Heibrock & Weis 1983). The transport of smoke in fires is strongly influenced by a number of factors, such as the geometry of the room, ceiling shape and height, fire load and burning characteristics of the contents, possible obstructions, the prevailing patterns of air currents within the protected space, etc. Especially important is the choice of detector location and spacing in enclosures with forced ventilation. Improperly located detectors in such spaces may not detect smoke at all owing to smoke dilution by clean air from a ventilation air inlet. A favorable location would be in the vicinity of the air return because of the greater probability that smoke will reach the detector and because the air velocity at the return is frequently lower.

5.1.3 Flame detectors

Operational principle

Flame detectors respond to the invisible portion of the radiation spectrum, either the ultraviolet (UV), or the near infrared (IR), emanating from flames (Middleton 1983). In comparison with the other types of detectors they display the fastest reaction times.

UV detectors use the radiation range 185–260 nm. As sensing elements, silicon carbide or aluminium nitride or, alternatively, a gas-filled tube can be used (Zimmerman 1986b). Besides flames, other likely sources of UV radiation are sunlight, atmospheric lightning and electric arcs from welding and cutting operations. Interference from sunlight is negligible since the Earth's atmosphere, owing to the shielding effect of the ozone layer, absorbs UV radiation below 280 nm. Therefore UV detectors can be used both indoors and in the open. However, the other two sources of UV radiation (lightning an cutting and welding) if present may cause false alarms. Another frequent problem with UV detectors is a gradual reduction in sensitivity due to the fouling of the optics by dust, smoke, soot, grease, oil, etc. In principle, UV detectors could be used in almost all cases where a flaming combustion in a protected space is expected. In practice, however, they are not employed very extensively, as IR detectors have turned out to possess a number of advantages. A notable exception is the use of the UV detector for the detection of flames from burning hydrogen which radiate intensively in the UV region.

IR detectors are widely employed in identifying the early stages of hydrocarbon fires. IR spectra of such fires exhibit a distinct band at 4400 nm, attributable to the stretching of the bond between oxygen and carbon in the molecule of CO_2 excited by the radiation from the flames. However, other sources of heat, such as sunlight, and heat from electrical motors, also absorb at or near this wavelength (Fig. 5.4). The

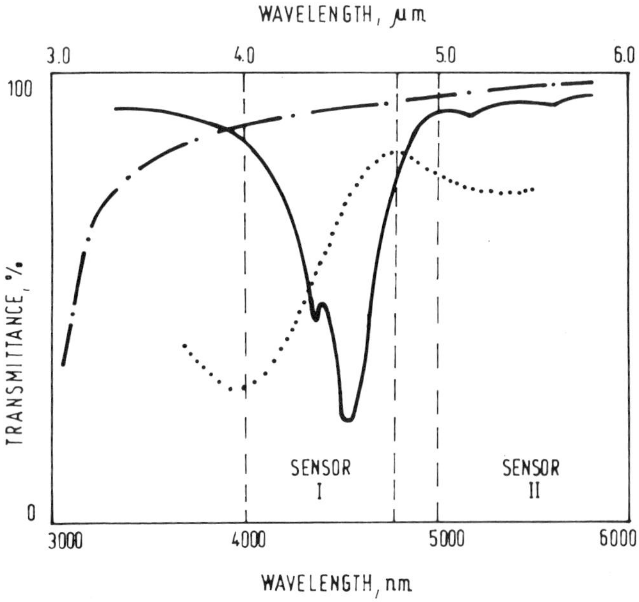

Fig. 5.4 — IR spectra of flame (——), artificial light (—·—), and sunlight (······) in the region of response of sensor of sensor I and sensor II.

selectivity of the detector to CO_2 is achieved by combining a selective filter and two pyroelectric sensors made of $LiTaO_3$ that respond at different wavelengths (Peissard 1985). The first sensor responds to the CO_2 band only, while the second is sensitive to unwanted IR radiation in the region 5000 to 6000 nm. Signals from the two detectors are compared and evaluated by an electronic circuit. In the case of fire, IR radiation from hot CO_2 significantly increases the signal from the first detector. The response sensitivity as well as the response time of IR detectors can be adjusted in several stages. They normally work in a pulse regime. The slow reduction in signal intensity, caused by fouling of the optics and aging, can be, to a degree, automatically compensated for by the use of a built-in microprocessor.

Application
When the protected area has a potential for the outbreak of a flaming fire which must be detected very quickly then flame detectors are the correct choice. Flame detectors are therefore installed mostly at highly hazardous areas, such as loading platforms and storage bays for flammable liquids, plastics, natural gas, certain chemicals and other products, the molecules of which contain a large proportion of carbon atoms. 'Explosion-proof' designs of these detectors are also available. Since the sensor must 'see' the flame and not the other fire products, the usual position of the detector is on walls (aside) rather than on the ceiling (directly above the fire). This also precludes any blocking of the detectors by objects placed in their angle of vision. In comparison with other detectors they can cover a wider area. Modern versions of IR detectors are insensitive to harsh ambient conditions, sunlight, and artificial lights, such as from halogen lights, or the light from cutting and welding operations due to the already mentioned method of dual-criteria evaluation of incoming radiation. This mean that the rate of false alarms is relatively small.

5.1.4 Gas-sensing detectors
Fires produce 'fire gases', which include carbon monoxide, carbon dioxide, hydrogen sulfide, and many other combustion products, often in trace concentrations. They are liberated in the early phase of fire growth, together with smoke, and reach their highest levels of concentration before the flaming phase. Chemically they are amenable to oxidation or reduction reactions, a fact which can be employed for the detection of their presence. Gas-sensing detectors contain a sensor which can be either a semiconductor or a catalytic element (Zimmerman 1986a). The former changes its conductivity following oxidation or reduction of fire gases, while the latter catalyzes oxidation of combustible gases.

Although in principle there should not be a serious reason why gas-sensing detectors could not be used on equal terms with other types of fire detectors, they are not often used because of the availability of the cheaper and more versatile smoke detectors. Instead, gas detectors are used more often in alarm systems which are there to detect gas or vapor hazards in enclosures at concentrations much below lower explosive limits.

5.2 AUTOMATIC EXPLOSION DETECTORS

Protection of spaces in which explosive combustion of fuel–air mixture may occur can be achieved by employing automatic explosion suppression systems (see subsection 7.4.4). A key component of such systems is an explosion detector. Explosion detectors differ from fire detectors in that they usually send a signal first to an alarm system (except heat detectors in sprinkler heads). Automatic explosion detectors are invariably coupled directly to suppression equipment because of the very short time (of the order of milliseconds) that elapses before the destructive effects of explosion are developed. The principles of an automatic explosion suppression system are shown in Fig. 5.5.

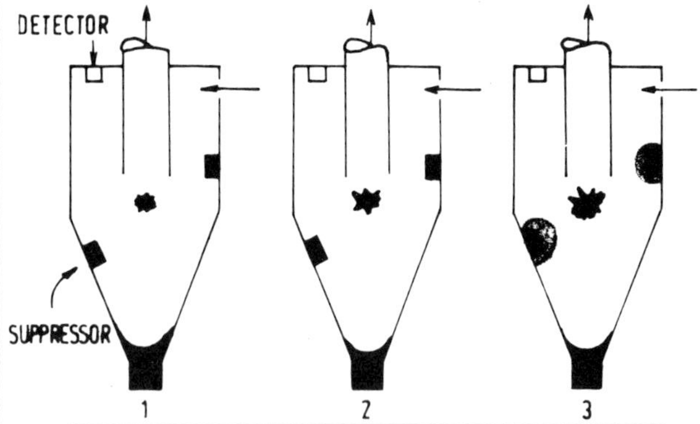

No.	TIME, ms	PRESSURE, kPa	ACTION
1	0.00	0.0	IGNITION BEGINS
2	0.25	1.9	DETECTOR RESPONDS
3	0.30	3.2	SUPPRESSION STARTS
4	0.35	6.0	SUPPRESSION CONTINUES
5	0.40	8.6	SUPPRESSION CONTINUES
6	0.50	15.9	SUPPRESSION COMLETED

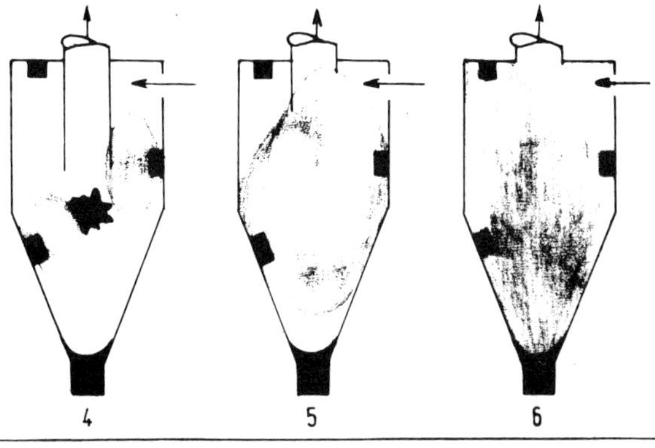

Fig. 5.5 — Principles of an automatic explosion suppression system.

In section 4.5 it was shown that the principal parameters that characterize an explosion are the maximum pressure developed and the rate of pressure rise. The same parameters are exploited in the design and operation of an important type of explosion detector. Explosion detectors which respond to explosion pressure are very sensitive pressure transducer devices (Peterson 1986). They contain a low inertia diaphragm capable of sensing pressure levels as low as 1.7 kPa. When in the early phase of explosion growth the pressure reaches a predetermined level the detector responds and activates the suppression equipment.

If normal variations of operational parameters of a process include large pressure fluctuations, then it may be more convenient to use a detector that senses the rate of pressure rise, rather than the absolute pressure level. These detectors respond when the rate of pressure rise exceeds a preset value (usually 35–70 kPa/s) (Peterson 1986).

Another explosion phenomenon which can be employed for the design of an explosion detector is the radiation generated at the incipient stage of an explosive combustion. The radiation can be detected by using either UV or IR fire detectors described in section 5.1.3. Radiation detectors are used in those cases where the principal characteristic of the explosion is not a rapid pressure buildup (for example, explosions in unconfined spaces). In comparison with explosion detectors, which respond to pressure, radiation detectors, being more sophisticated, cost much more.

Application

Explosion detectors, as integral parts of explosion suppression systems, are used almost exclusively in industrial settings involving such hazardous process operations as size reduction processes, dust collection, spray painting, mechanical and pneumatical conveying, solvent recovery, etc. Before the selection and application of explosion detectors, the explosion characteristics of the particular fuel–air system should be known. Explosion detectors are located at those places where the formation and ignition of an explosive fuel–air mixture are more likely.

The potential number of explosion suppression systems involving the use of automatic explosion detectors far exceeds the number of systems actually installed and operated. Explosion venting, as an alternative method of explosion protection, is still used much more frequently (see subsection 7.4.4).

5.3 PROTECTIVE SIGNALING SYSTEMS

Fire detectors themselves cannot function satisfactorily unless they are part of a protective signaling system. There are many types of protective signaling systems, ranging from very simple to highly complex ones, which integrate several protective functions (fire alarm and control, burglar alarm, door entry control, etc.) (Zimmerman 1986b). However, all systems, regardless of their type and degree of complexity, have some common basic features. These include a means of fire detection, signal indication and control, equipment for alarm initiation and reception, and power supply.

In addition to automatic fire detectors a fire can be detected and signaled by either a manual alarm box activated by a guard or other employee, or a sprinkler system waterflow alarm and signaling service.

Alarm, supervisory and trouble signals from signaling equipment are received, recorded, maintained and supervised at a central control unit. The central control unit, on verification of the fire alarm, transmits it to the local or public fire department for corrective action.

The power supply for protective signaling systems should be obtained from two independent sources of electric power of adequate quality and capacity. The primary (main) power is obtained from the commercial light and power service or from an engine-driven generator, while storage batteries or an engine-driven generator may serve as sources of secondary (standby) supply. Fire signaling systems must have provisions for the indication of interruption of the main power supply, an indicator of the battery status, and an indicator of the detector fault.

The working principles of modern fire signaling systems remain unchanged; things that do change are the base elements and the system construction. Thus the size and mass of systems are constantly being reduced, electric power requirements are lowered, false alarm rates are decreased, while the reliability of systems and their working lives are being increased. It is expected that future protective signaling systems will also be able to give quantitative information on the parameters they control, such as smoke density, level of temperature, intensity of UV or IR radiation, etc. (Rohrbacher 1983).

Modern fire signaling and protective systems should be able to fulfil several important functions, for example: (a) determination of the correct location of the place from which a protective signal originates, (b) information on the status change of each detector, (c) protection against false alarms and (d) the reduction of faults to a minimum (*Schadenprisma* 1983). They should also be highly reliable. If for any reason (malfunction, repair, break) the system is not working for a short period, say two days a year, or *ca.* 10 minutes a day, the reliability of the system is reduced from 100% to 99.5%. The microprocessor in the central station must provide for a consistent testing of detectors connected to the transmission channel and obtain information on the status of each detector. The parameters of each detector are kept in the memory of the computer. Checking the status of all the detectors can be done in a very short time.

A still unresolved serious drawback of all the signaling systems is a high false alarm rate. Some fire statistics put it as high as 98% (*Fire Prot.* 1985a). False alarms can be caused by environmental factors such as high temperatures of a nonfire origin, high humidity, direct sunshine, process dust, cigarette smoke, etc. It would seem reasonable that before setting the threshold of an automatic fire detection system, measurements of the protected atmosphere and microclimatic conditions should be performed over an interval of, say, one or two weeks. Modern systems, with built-in microprocessors, have somewhat reduced the frequency of false alarms, but have not yet been able to eliminate them. Microprocessors receive the signal, evaluate it and decide whether the alarm system should be actuated or not. The probability of false alarms can also be reduced by the use of systems which are activated only when the control station receives a fire alarm from at least two different fire detectors (for example from a smoke and heat detector). This is known as a 'voting system'.

Some fire detectors are very sensitive and show a high rate of false alarms. In such cases the central station may first retard the received signal and send a patrol to check the condition at the location from which the alarm was received. The alarm system is then activated only when the patrol confirms the fire.

The high false alarm rate has caused firemen to be suspicious of the alarms received from automatic fire detecting and signaling systems. Yet, the practice of fire detection and signaling reveals that the overwhelming majority of false alarms are merely due to improper exploitation of the protective systems.

A further undesirable feature of modern systems is a high diversity of indicating and protective systems.

Depending on the size and value of the protected premises, protective signaling systems can be divided into local, auxiliary, remote, proprietary or central station signaling systems (Zimmerman 1986b). Basic features of each of these systems will be briefly reviewed.

5.3.1 Local protective signaling systems

Local protective signaling systems are the simplest systems intended for protection of life and property at a particular locality, for example a building, or a piece of processing equipment (NFPA 72A 1985). A fire signal initiated by a fire detector(s), or by manual operation of a fire alarm box, terminates at the protected place, which means that the fire brigade must be notified separately by some means (usually by telephone, or by radio). In the meantime, after the initiation of an audible and/or visual alarm, the fire is presumed to be tackled by manual or automatic fire extinguishing equipment, while the occupants are evacuated from the building (Fig. 5.6).

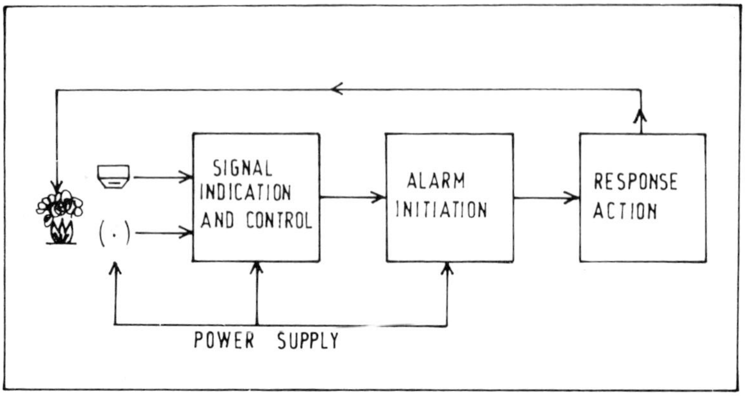

Fig. 5.6 — Basic components of a local protective signalling system.

In the case of a circuit break or grounding occurring in the detector or associated wiring, a trouble signal, distinctive from the fire alarm signal, is sounded.

Local protective signaling systems can be designed for other purposes, such as automatic detection of abnormal conditions in industrial processes or extinguishing systems which could be hazardous to life or property.

5.3.2 Auxiliary protective signaling systems

Auxiliary protective systems are designed on much the same principles as local ones, except that there is a probabality of summoning the fire department by use of a municipal fire alarm system (Fig. 5.7). They provide protection to either an individual building or a group of buildings of a single occupancy (NFPA 72B 1986).

5.3.3 Remote station protective signaling systems

In remote station protective signaling systems, compared with local and auxiliary protective signaling systems, fire signals initiated by fire detectors, manual fire alarm

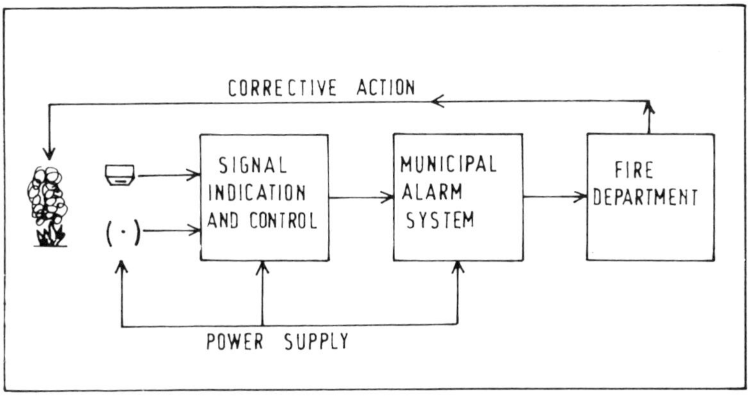

Fig. 5.7 — Basic components of an auxiliary protective signaling system.

box or sprinkler water flow alarm system, as well as supervisory and trouble signals, are transmitted first to a remote station (Fig. 5.8). The remote station is manned at all

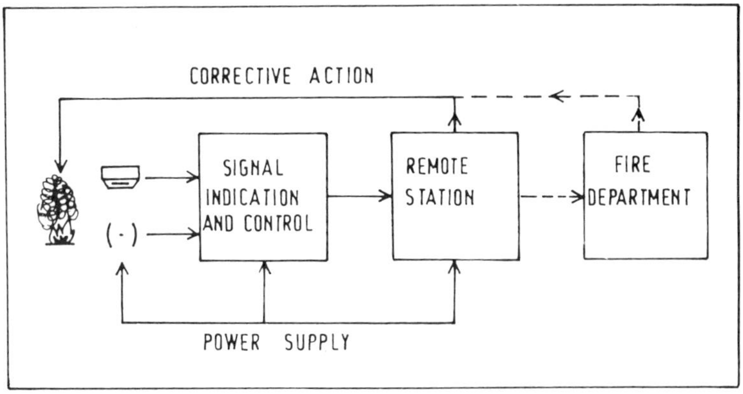

Fig. 5.8 — Basic components of a remote station protective signaling system.

times by competent and trained personnel who supervise the circuits and investigate signals (NFPA 72C 1986). The station itself may be a police or fire station or may be connected to them by a transmission system (a dedicated circuit for voice or coded signal communication, a one-way telephone, or a radio system). Signals received at the remote station each initiate a distinctive automatic visual and audible indication, and are manually or automatically recorded and are logged by date and time of the appearance of the alarm or abnormal conditions.

5.3.4 Proprietary protective signaling systems

Proprietary protective signaling systems (NFPA 72D 1986) are designed to protect a property through a central station located at that property (Fig. 5.9). The station is attended by trained and competent personnel at all times. The operators at the station constantly monitor received signals, operate the system and when necessary take the required action. When a fire signal is verified it is transmitted to the fire department by wire, radio or equivalent means. Modern proprietary protective systems employ computerized controls and have a signal multiplexing capability. Thus, fire detection and alarming is just one, ableit the most important, of a number of other functions provided.

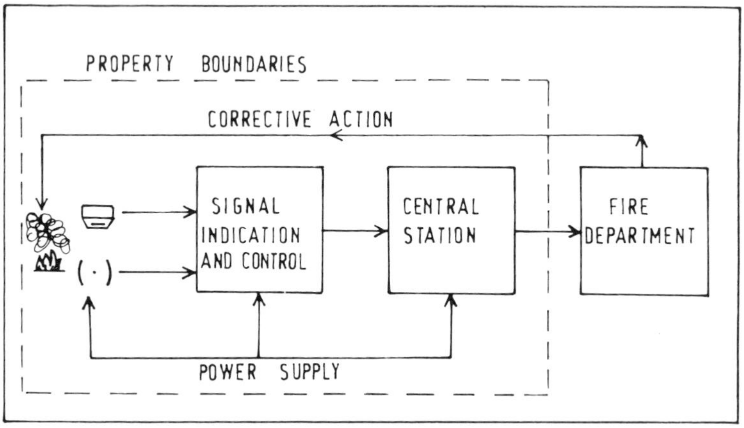

Fig. 5.9 — Basic components of a proprietary protective signaling system.

5.3.5 Central station protective signaling systems

These systems are designed similarly to the proprietary protective signaling systems. The principal difference is that the central station is located away from the protected property, sometimes at a considerable distance. Alarm and supervisory signals from protected buildings are first collected at an unattended satellite station from which they are transmitted to the central station through an exterior multiplexing communication channel (Fig. 5.10). At the central station the operations of the circuits and devices are automatically signaled, recorded, maintained and supervised (NFPA 71 1985). Operators at the station continuously monitor the type of signal (whether an alarm, supervisory, delinquency or trouble), its condition (whether an initiation of an alarm, restoration or return to normal) and its location. Fire alarm signals, readily indentifiable as to location of origin, are retransmitted from the central station to the fire department by at least two independent channels. In the meantime a patrol is promptly dispatched to the protected property for alarm verification and, if necessary, detector and equipment resetting.

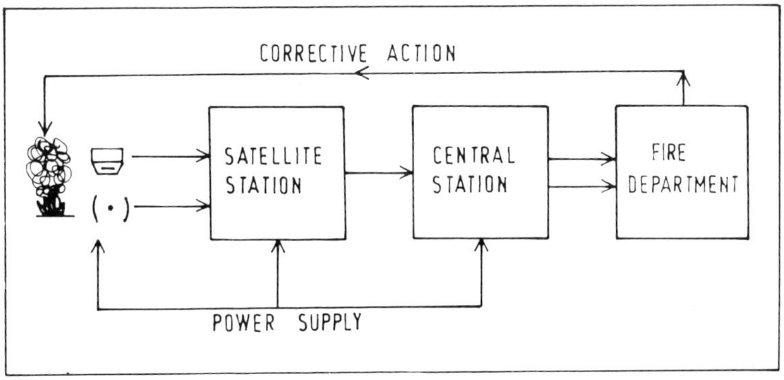

Fig. 5.10 — Basic components of a central station signaling system.

The central station may be supplemented by devices for automatic fire protection control, such as for closing fire doors, smoke evacuation, activation of fixed extinguishing systems, control of lifts, etc.

6

Effects of fires and explosions

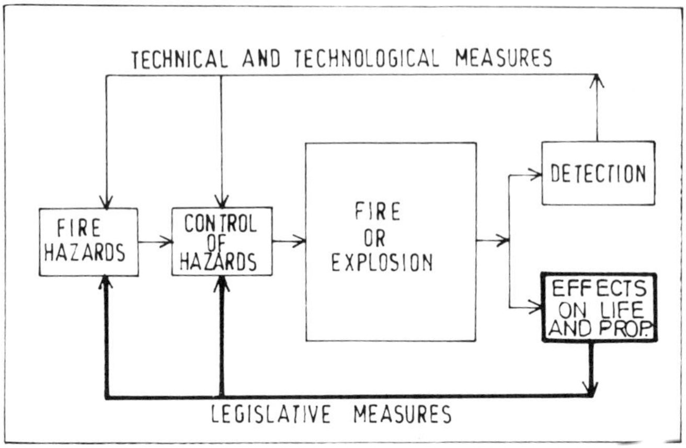

Fires or explosions may be portrayed in the form of an uncontrolled chemical oxidation reaction written in block form in equation (6.1)

$$\boxed{\text{Combustibles}} + \boxed{\text{Oxidant}} \xrightarrow[\Delta]{} \begin{matrix}\text{Smoke}\\ \text{Heat}\\ \text{Gases}\\ \text{Flame}\\ \text{Pressure}\\ \\ \text{Fire or Explosion}\end{matrix} \qquad (6.1)$$

What will result from the above reaction is determined mainly by its kinetics. Extremely rapid oxidation, accompanied by pressure rise and chemical transformation of the system into mechanical work, is pertinent to explosions, while fires involve a more moderate reaction rate.

However, regardless of the rate of oxidation, products of the reaction in either case can be classified into several distinct categories, among which the most important are *smoke*, *gases*, *heat*, *flame* and *pressure*. Each of these products exerts specific effects on people and their environment. The effects may thus be divided according to the products of the oxidation reaction, or to recipients of fire and explosion impact.

6.1 OVERALL EFFECTS

6.1.1 Effects on people

Each year throughout the world, fires and explosions take a hcavy human toll. Thousands of people are killed, and many more are injured, often permanently. Recent fire statistics for 12 countries (*Fire* 1986c) reveal that on average the number of fire deaths per 100 000 of population varies between 0.54 (for Switzerland) and 2.50 (for the USA) (Table 6.1).

Table 6.1 — International fire deaths, 1983

Country	Deaths per 100 000 population
Switzerland	0.54
Netherlands	0.59
Austria	0.96
Yugoslavia*	1.15
Spain	1.20
New Zealand	1.25
Denmark	1.37
Norway	1.45
Japan	1.56
Sweden	1.68
France	1.70
Finland	1.92
United Kingdom	2.02
United States†	2.50

*Data for Yugoslavia from the official 1984 Report.
†USA data from Cote (1968).

The world's leading rates are firmly held by the USA and Canada in that order (Cote 1986) since these two countries annually experience by far the largest number of fires (Maguire 1984). A generally apathetic attitude to fire prevention in these two countries (*Fire Contr. Dig.* 1984a) and an inferior fire protection organization in comparison with European countries (Schaenman 1983), coupled with the considerable economic impact of fire protection, seem to intensify such a high fire death rate.

The chances of being killed by fire have been estimated to be 1:60 000 per year (Rasbash 1984).

The lowest fire death rate is in the 15–35 age group, as this group is able to evacuate more rapidly in the event of fire (Gormsen *et al.* 1984). Younger people (children up to five) and older people (over 65) are the most likely victims, as fire deaths of these groups are disproportionally higher (Maguire 1984), since they spend most of their time at home. Consequently, most fatal fires occur in residences (Ellis & Biegel 1982). Although one- and two-family dwellings and mobile homes account for 64.2% of all residential-type occupancies (Ellis & Biegel 1982), fire risks in high-rise buildings are the most severe (Müller 1983), and often produce multiple deaths. It has been found that one in every 100 fires in residential occupancies causes death (Maguire 1984). Deaths in sprinkler-protected buildings are strikingly lower (Hagglund 1983), which shows that the main function of sprinklers — the protection of lives — is amply proven.

There have been numerous attempts to correlate fire deaths with a number of different factors. Thus, especially high positive correlation is found with alcohol consumption (Gormsen *et al.* 1984, Banks & Montgonery 1983), and smoking (Maguire 1984, Banks & Montgomery 1983, McGuire 1983). Smoking has been blamed as the principal cause of multiple-death fires in the USA (*Fire Prev.* 1983). Other statistically significant correlations have been found with cold climates (a positive correlation) and the number of fire fighters per unit of population (a negative correlation) (Banks & Montgomery 1983).

Taking the generally higher fire loads in industry one would expect a relatively large number of fire-related deaths in this category of occupancy. However, the fire death toll in industry is surprisingly low, since the inherent fire dangers are better controlled (*Fire Prev.* 1984).

The large fire death rate has led to pronounced research, technical and legal interest as to its cause. The results have led to a significantly reduced number of fire-related deaths (up to 25%) in the last 10 years in the US alone) (Maguire 1984, McGarry 1983).

Fire fighting is a dangerous profession. Thus, each year a large number of fire fighters die in the line of duty (*Fire Contr. Dig.* 1984b). The leading cause of deaths is not smoke intoxication, or burns, but heart-related diseases (*Fire Contr. Dig.* 1984b, Clark 1984) brought about by high stress (Hildebrand 1984). Especially vulnerable are older firemen for whom heart attacks are the leading cause of death (Washborn *et al.* 1986). Fire fighters often die in groups because of explosions during the intervention, or in fighting forest fires and in transportation-related incidents (Hall 1984).

Fires and explosions also cause a great number of injuries. Fire injuries are defined as the effects of fires or explosions on people who then require medical attention and treatment (NFPA 901). Fire injuries exceed the number of deaths many times (Ellis & Biegel 1982, Einhorn 1983). Injured people always experience pain and often need long hospitalization, further enhancing overall fire costs and effects.

Physiological changes in fire fighters due to the weight of personal equipment under hot environmental conditions are reflected in increased heart rhythm, increased concentration of noradrenaline and increased sweating (Nodziri & Marujama 1984). These conditions increase air consumption and energy expenditure during fire operation (Harikae & Morita 1984).

6.1.2 Effects on property

In addition to human losses, fires and explosions cause tremendous wastage of property. Property losses in industrialized countries vary between 1.0 and 1.5% of gross national product (Rasbash 1984). About one fifth of this value comprises direct property losses, while the remainder is made up of the costs of conducting fire prevention, of maintaining fire departments, of conducting the fire insurance business and of the necessary organizational measures (Rashbash 1984).

Indirect losses from fires and explosions are hard to assess. They often cannot be measured in monetary terms, such as loss of credit standing, loss of trained personnel, loss of customer confidence, etc. The World Fire Statistics Centre lists seven key parameters which indicate fire losses (*Fire Int.* 1983), and calls for a uniform reporting of fire losses using these parameters as a base.

A large percentage (*ca.* 70%) of total fire losses occurs in high-valued industrial and commercial properties (*Fire Protection Handbook* 1977), with most fires occurring in storage occupancies (*Fire Prot.* 1985b). In an analysis of the role of fire defence within UK industry it has been estimated that fire losses in industry amount to £6000 per minute if fires are not extinguished (Durrant 1985).

6.2 EFFECTS OF FIRE AND EXPLOSION PRODUCTS

6.2.1 Effects of smoke

Smoke in fires appears as a result of nonstoichiometric combustion of fuels. In addition to the final oxidation products, CO_2 and H_2O, combustion products contain a number of gases and partially oxidized and reduced compounds, such as methane, CH_4, methanol, CH_3OH, formaldehyde, HCHO, formic acid, HCOOH, acetic acid, CH_3COOH, as well as droplets of flammable tars, condensed vapors and very fine solid particles (Hurst & Jones 1985). The presence of these products produces a visible appearance of combustion products known as smoke. Smoke in fires is evolved practically at all temperatures (Parry 1983). The physical conditions of combustion, such as the combustion rate, the combustion mode and the temperature, have more influence on the smoke composition than does the kind of burning material (Gad & Smith 1983, Fardell *et al.* 1986).

The characteristics of smoke that are most dangerous to people are its toxicity, color and density.

Fire statistics and research reveal that most fire deaths are caused by smoke inhalation (Gormsen *et al.* 1984, Ellis & Biegel 1982, Taylor 1984, *Fire* 1985). Modern materials in fires release more toxic compounds per unit weight than do the traditional ones, and although the smoke composition has remained fairly constant, the rate of fire spread and the emission of smoke have increased. These two factors are the leading causes of the increased death rate in fires (*Fire J.* 1984) Inhaled smoke causes irritation and damage to the respiratory system, while smoke in air affects the eyes, inducing tears, and in more serious cases even injury to the eyes.

Smoke color varies, depending on the material being burned, from light blue in the case of good combustion to heavy black during the combustion of some high molecular weight hydrocarbons. Dark-colored smoke significantly reduces visibility, obscuring exit signs, and induces panic in people in fires (Benjamin 1984).

Another important smoke characteristic from a safety point of view is its density. It is well known that dense, copious smoke obscures visibility and threatens the lives of both the people being rescued and the fire fighters. The visibility reduction depends not only on the composition and concentration of smoke but also on the radius of the smoke particles, on the nature of the light, as well as on the psychophysical state of the observer (Butcher & Parnell 1983).

6.2.2 Effects of fire gases

The principal effect of fire gases is in their toxicity when inhaled, sometimes even in very low concentrations. The toxicity is enhanced since the inhaled gases are generally at a high temperature and often oxygen deficient. However, in case of death it is very difficult to pinpoint a single gas that is presumably responsible for the tragic outcome (*Fire Protection Handbook* 1977). Fire gases appear in mixtures and the possibility of sinergism is always present. The situation is further complicated owing to the nonuniformity in medical appraches to the determination of the causes of death (Maguire 1984). In addition, the great majority of toxicity tests refer to laboratory conditions, which are hardly able to simulate real fire scenarios (Clarke 1983). Even the tests are of dubious value. It has been shown that results obtained by a particular test are poorly reproducible even when performed in the same laboratory (Huggett 1984). Furthermore, the questions of toxicity criteria and the validity of the extrapolation of toxicity data obtained with animals to people are still unresolved, which makes the problems due to toxicity of fire gases very complex. Therefore a great deal of research concerning the toxicity of smoke and fire gases is being undertaken in many countries (Queen 1983).

However, what is known for certain, on the basis of numerous biological and chemical experiments, is that many fire gases, taken individually, are toxic to humans. The toxicity depends on several factors, of which the concentration of the gas in air, the duration of the exposure and the physical condition of the individual are the most important. The toxic characteristics of some gases that most prevail in fires are described below.

Carbon-containing gases

Carbon monoxide. A great majority of combustibles contain carbon atoms in their structures. During fires, carbon is first oxidized to carbon monoxide:

$$CO + 1/2\,O_2 \rightarrow CO \qquad \Delta G = -137.3\ \text{kJ} \tag{6.2}$$

which is a chemically inert gas containing a strong triple bond between C and O. The subsequent formation of carbon dioxide, CO_2:

$$CO + 1/2\,O_2 \rightarrow CO_2 \qquad \Delta G = -257.2\ \text{kJ} \tag{6.3}$$

is considerably slower owing to the large activation energy of the CO molecule. Moreover, in most fires, especially in the early stages, there is a deficiency in optimal oxygen concentration, causing conversion of CO to CO_2 to be incomplete with large amounts of CO remaining unreacted.

It has been known for a long time that CO is the most hazardous gas in a fire atmosphere. Being a colourless, odorless and tasteless compound, CO is a perfidious gas. Although it is not the most toxic of fire gases its abundance makes it the chief danger.

When inhaled, CO interferes with the transfer of oxygen throughout the human body by forming a coordination complex with hemoglobin (Hb) in the red blood cells:

$$CO + HbO_2 \rightleftarrows HbCO + O_2 \tag{6.4}$$

$$K_{eq} = \frac{[HbCO][O_2]}{[CO][HbO_2]} = 210 \tag{6.5}$$

displacing oxygen from hemoglobin and preventing it from being transported through the bloodstream. The rather large value of the equilibrium constant for the reaction (6.5) means that CO when present in even very small concentrations is able to compete successfully with ordinary concentrtations of oxygen. The narcotic effects of CO, reflected in the impairment of physical abilities, are explained by this deprivation of various vital organs, especially the brain, of the needed oxygen (US EPA 1970).

The formation of carboxyhemoglobin in blood is a reversible process (Connor 1984). Levels of HbCO in the blood take twice as long in smokers than in nonsmokers to return to normal (0.5–2%) (Connor 1984).

The effects of CO depend on a number of factors, of which physical activity and length of exposure are again the most important. CO poisoning is more rapid with people involved in some kind of heavy work than with people at rest (Wolf 1971). Other factors of importance, which increase the effects of CO, are high temperature, decrease in O_2 concentration, previous chronic diseases (especially heart and lung-related), presence of alcohol and narcotics in blood, the feeling of anxiety and fear (Einhorn 1983).

The observable effects of CO (headache, nausea, sweating) are experienced at rather low concentrations (about 100 ppm) (Cagliostro & Islas 1982). At higher levels (about 500 ppm) there is loss of consciousness and at still higher levels approaching a few percent, death may result after just a few breaths (Beech 1983). A 40% saturation of blood by carboxyhemoglobin leads to loss of consciousness, while about 50% saturation is found in the blood of CO victims (Packham 1984). An empirical rule, applicable for the range of CO concentrations normally formed in fires, states that dangerous effects of CO can be expected in any situation in which the product of CO concentration (in ppm) and the length of exposure (in minutes) exceeds *ca.* 35 000 ppm/min (Packham 1984).

Carbon dioxide, CO_2. Carbon dioxide is the final product of combustion of materials containing carbon:

$$C + 1/2\, O_2 \rightarrow CO_2 \tag{6.6}$$

The reaction actually occurs in two stages, as shown in equations (6.2) and (6.3). Concentrations of CO_2 in fires normally reach 12–13% (Fardell *et al.* 1986).

Carbon dioxide itself is not considered poisonous. Increased levels in the air mean that concentrations of other gases normally found in air are altered. Thus O_2 concentration is reduced, causing an increase in the rate of breathing and the intake of smoke and other gases that may be present. The effects of CO_2 on human health are therefore more indirect than direct.

Carbon dioxide is widely used in fire protection systems as a gas that effectively extinguishes fires in closed or semiclosed spaces by oxygen dilution. There have been instances where people incapable of escaping confined spaces in fires have become victims of CO_2 owing to suffocation.

Nitrogen-containing gases

Nitrogen oxides, NO_x. The combustion of materials in air always produces some oxides of nitrogen. They are derived both from the combustion of fuel and from the high-temperature reactions of oxygen and nitrogen in air. Although there are at least eight known nitrogen oxides (Cotton & Wilkinson 1972), two of them, nitrogen oxide, NO, and nitrogen dioxide, NO_2, are present in the largest amounts. Since most methods of chemical analysis measure their combined total concentration, these two oxides are usually treated together and labeled NO_x ($x=1,2$). In any fire, NO is by far the more prevalent, with only *ca.* 5% of NO_x being NO_2 (US EPA 1982). However, NO is continuously oxidized to NO_2, so after some time the concentration of NO_2 may be quite high (US EPA 1982).

With respect to reactivity and life safety, NO_2 is much more reactive and extremely toxic. Concentrations that could be safely breathed for a short time are very low, in the low ppm range (Butcher & Parnell 1983). If the materials being burned are rich in nitrogen, as in some synthetics (e.g. polyurethanes), or in inorganic nitrates, concentrations of NO_2 may be high enough for a characteristic reddish-brown color to be easily perceived. Toxicological effects of NO_2 are often delayed, making misdiagnosis of injury or death a real possibility.

Hydrogen cyanide, HCN. Hydrogen cyanide is produced as a result of combustion of nitrogen-containing materials under reduced conditions (Levin *et al.* 1985). In low concentration it smells like bitter almond, while in higher proportions it has a very unpleasant odor, albeit an unreliable one. Many people soon become quite insensitive to its odor, or the odor may be masked by other fire odors.

The highly toxic nature of HCN results more from its extremely fast action than from its minimum lethal dose. Hydrogen cyanide deaths occur swiftly owing to anoxia and the interference with a specific enzymatic reaction in the body, often followed by circulatory collapse. The empirical rule of the product of concentration and exposure time mentioned earlier (Packham 1984) as a practical threshold for hazardous effects of a fire gas is significantly lower (*ca.* 1500 ppm/min) than that for carbon monoxide.

Amonia, NH_3. Smoldering fires of many synthetic materials often produce ammonia, NH_3. In well ventilated fires, ammonia and other reduced nitrogen compounds are short-lived and are rapidly oxidized to nitrogen oxides.

Ammonia is a toxic gas with a characteristic pungent odor which is highly irritating and repulsive to people. People who experience ammonia-containing atmospheres suffer severe irritation of the eyes and the respiratory organs. If the ammonia concentration increases to over 200 ppm for a few minutes, serious injuries or death may result (Butcher & Parnell 1983).

Sulfur-containing gases

Hydrogen sulfide, H_2S. In smouldering combustion of materials containing sulfur in their structures (i.e. hides, hair, meat, wool, rubber), hydrogen sulfide, H_2S, is likely to evolve. Its presence is immediately recognized by a 'rotten egg' smell and taste which act as an efficient warning to people. If people do not escape quickly enough the sense of smell is rapidly depleted, leading the way to serious poisoning at very small concentrations (a few hundredths percent by volume). At somewhat larger concentrations (600 ppm), hydrogen sulfide acts as a nervous system depressant, causing an extremely rapid breathing rate followed by paralysis of the respiratory tract. Prolonged exposure to this concentration ends in death.

Sulfur dioxide, SO_2. Sulfur dioxide is the end product of oxidation of sulfur-containing materials, such as wool or rubber (if a very small concentration of SO_3 is neglected). It is very irritating to the eyes and the respiratory organs. The concentrations normally attained in fires are rather low. However, even these small concentrations are hazardous. The maximum allowed short-term exposure is estimated at about 500 ppm (Butcher & Parnell 1983).

Halogen-containing gases

Hydrogen chloride, HCl. Hydrogen chloride evolves on heating polymeric plastic materials, such as polyvinyl chloride (PVC). Since polyvinyl chloride is widely found in everyday life, hydrogen chloride is often produced in fires. Observable evolution of HCl from polyvinyl chloride under laboratory conditions begins at 520–570 K (Smith 1985) and is very fast at 720 K. Although one would expect combustion products of polyvinyl chloride to be considerably toxic in comparison with more traditional materials, such as cellulose or wood, polyvinyl chloride in fires is not more toxic (Hinderer 1984). Moreover, analysis of the literature data on the experimental determination of toxic products of thermal degradation and combustion of polyvinyl chloride suggests even lower toxicity of PVC-originated gases (Hinderer 1984).

Although 58% of the mass of PVC in fire should theoretically be lost as HCl, in practice, only one third of this value is found in fire gases (Brown & Martin 1985). The rest is lost, because of the reaction of HCl with water, by precipitation on walls and floors, and for other reasons. However, the remaining HCl can be fatal in low concentration (*ca.* 1500 ppm) (Butcher & Parnell 1983). Its pungent and irritating odor may seriously hurt firemen (Smith 1985) and its high reactivity leads to corrosion of metals.

Hydrogen fluoride, HF. Polymeric materials containing fluorine when involved in fires produce hydrogen fluoride, HF. The evolution of HF starts at *ca.* 570 K and is rapid at 770 K (Smith 1985). HF is a hundred times more toxic than HCl. Therefore the use of gas-masks and protective clothes is necessary when fighting fires in HCl- and HF-laden atmospheres.

Oxygen deficiency

Oxygen deficiency, together with the presence of CO_2 and CO, has been recognized as one of the leading causes of death in fires (Gormsen *et al.* 1984). Hot and predominantly reducing gases, coupled with oxygen consumption by combustibles, decrease the oxygen concentration normally found in air. The reduction from the normal 21% to 19% does not produced any visible distressful effects (Cagliostro & Islas 1982). A further decrease to around 16% increases the rate and depth of breathing (Beech 1983), while concentrations around 14% interfere with the normal evacuation procedure. Still lower concentrations (*ca.* 12%) produce serious incapacitation, followed by death if the concentration drops to about 6% for a few minutes (Beech 1983).

Other gases

In addition to CO_2, CO, nitrogen-, sulfur-, and halogen-containing gases as the principal fire gases, real fire atmospheres are loaded with a great number of other gases, present mostly in very low concentration (for example, polycyclic aromatic hydrocarbons, aldehydes, ketones) (Fardell *et al.* 1986, Crummett & Townsend 1984, Longwell 1982). The nature of these gases is dependent not only on the materials being burned, but also on the combustion mode, heating rates, oxygen concentration, air flow and exposure conditions. Although the low concentration of these gases and a generally short exposure to smoke probably preclude fatal effects, the possibility of synergism is ever present.

Recently, some very reactive species, 'free radicals', have been discovered in fire atmospheres. As to the effects on people, they act in much the same way as CO (*Fire Chief* 1985), i.e. blocking blood saturation with oxygen. It is assumed that on inhalation free radicals form certain peroxycompounds with lipids of lung tissue, making it incapable of gas absorption (*Rev. Techn. Feu* 1979). Blood analyses of deaths sometimes do not reveal formation of carboxyhemoglobin at lethal concentration levels, implying that death might have been caused as a result of the effects of free radicals. After breathing free-radical-containing smoke, affected people very often experience cardiac problems.

Despite their high reactivity and consequent short lifetime, free radicals may be present in fire atmospheres long enough to cause serious injuries to people. Since they begin to evolve at temperatures as low as 370 K in the first few minutes of fires their concentration may be substantial, exceeding three times the concentration of CO (*Fire Chief* 1985). Therefore firemen are advised not to enter objects under even the slightest smoke concentration without gas-masks, or some other means of air filtering (a piece of garment fabric, or a handkerchief will do).

6.2.3 Effects of heat and flame

During fires, enormous amounts of heat are often liberated. If for any reason (unconsciousness, drunkenness, physical or mental incapacity, drugs) people are not able to leave the fire atmosphere the effects of heat ranging from simple discomfort to death may develop. Elevated air temperatures can be tolerated to a certain degree, depending on air humidity, heat-shielding effects of clothing and physical activity (Philips 1980). The higher the moisture content of the air breathed, the lower is the maximum survivable breathing-level temperature. Since water vapor is one of the principal combustion products, and in many cases is added when used in fire fighting, a fire atmosphere is frequently moisture-laden. In such a situation skin perspiration is enhanced, leading to a rapid body dehydration. In experiments showing physiological changes of fire fighters by weight of personal equipment under hot environmental conditions (Nodziri & Marujama 1984) it was found that at 320 K and a relative humidity of 70% fire fighters lose up to 718 g of bodily fluid in eight minutes. The loss of bodily fluids shoud be balanced, if possible, by drinking water solutions of some salts ($NaHCO_3$, NaCl), or tap water (Newman 1982).

Water-deficient blood becomes denser, slowing the blood circulation (Sekine *et al.* 1982) and causing an increased heart rate. Physically fit men are able to endure air temperatures of 353–373 K for 10 minutes without serious consequences (*Rev. Techn. Feu* 1979). At slightly higher temperatures (386–392 K) breathing and heart rate are, however, so increased that after a short time heat exhaustion and total collapse set in (*Rev. Techn. Feu* 1979). Levels of tolerance to different temperatures are shown in Table 6.2 (Philips 1980).

Table 6.2 — Human tolerance to high temperatures

Air temperature K	Tolerance min
338	Can be tolerated for some time
>368	Sharply decreased
393	15
418	<5
448	Irreversible skin damage in less than one minute

The role of heat-shielding clothing is obvious. For effective protection against high temperatures, heat-resistive fabrics, which at the same time are capable of removing water extracted from the body, should be used. Ordinary, everyday clothes, of course, are not made of such a fabric, which means that they barely provide any heat protection. Moreover, the widespread use of relatively easily flammable synthetics actually increases the chances of being burned by fire.

Skin can be damaged when exposed to just minute amounts of heat flux. With some people even sunbathing can cause a severe heat shock. The threshold of human tolerance to heat radiation is sharply reduced in the interval 1.7–2.2 kW/m^2 (Hasemi & Shigekawa 1981). For instance, a heat intensity of 1.7 kW/m^2 can be tolerated for 30 minutes, while at 2.2 kW/m^2 heat can be endured for only five minutes (Hasemi & Shigekawa 1981). Exposure to still larger heat flux causes severe skin burns. A one second-exposure to 20 kW/m^2 leads to skin redness, while radiation of 28 kW/m^2 brings about skin burns (Bull & Lawrence 1979). Heat flux of 37 kW/m^2 seriously damages fabrics.

People are instinctively afraid of fires, so burns from flames can be caused only if there are no means of egress from fire, or people are incapacitated. Consequently, small children and very old people are the most likely victims (Feek *et àl.* 1979). People overwhelmed by fires, if they survive, usually end up with bad burns. Even then, their lives are endangered because of possible infection.

Burns may be brougnt about either by direct contact with flames, or by heat radiated from flames. When the surface temperature of the human skin increases to 44°C the outer layer of the skin becomes damaged, while at a temperature of 350 K skin tissue is immediately destroyed (*Rev. Techn. Feu* 1979). These extremes, together with intermediate levels of skin burns, are commonly classified as first-, second- and third-degree burns. Heat damage to a relatively small skin area can be very painful, and if a larger area is involved (50% for young people, 20% for old people) the death rate can reach 50% (Bull & Lawrence 1970). For comparison, 1% of the human skin area corresponds approximately to the area of the palm of the hand.

The probability of skin burns depends, in addition to heat intensity, on the type and temperature of materials being touched. At the same temperature of different hot objects the effective temperature at the point of contact between the palm and the object depends on its heat conductivity. Thus, if cylinders made up of steel, glass, wood and cork are heated to 373 K, and are touched by the palm, the effective temperature at the point of contact will be 371, 355, 338 and 319 K, respectively (Bull & Lawrence 1979).

6.2.4 Effects of high pressure

Explosions also exert heavy material damage and often cause multiple deaths. Of the combustion products of an explosive reaction the most important are overpressure and heat, since they contribute the largest share of the total explosion damage.

The effects of military and dense explosives on people and the environment are well known. The overpressures liberated are far greater than those from rarefied chemical explosions. Hence, the formation of craters in the ground, shattered buildings and heavy human losses are the direct, more visible, effects of explosions of dense explosives on the surface of the ground. Although peak overpressures of rarefied chemical explosions reach a maximum at a few thousand kPa, effects of such explosions can be very serious because of high explosion impulse (Marshall 1987).

Levels of explosion overpressure can be estimated by inspecting the kind and magnitude of effects on people and material goods. Some of the effects are shown in Table 6.3 (Cruice 1986).

The heat liberated during an explosion is of such intensity that flammable and

Table 6.3 — Effects of shock waves from chemical rarefied explosions

Peak overpressure, (kPa)	Effects
3.5–7	Shattering of glass windows. Personnel knocked down
14–21	Shattering of nonreinforced concrete wall (200 mm thick)
21–28	Collapse of oil storage tanks
48	Overturning of loaded rail cars. Eardrum rupture and some lung damage

combustible materials are frequently ignited. Ensuing fires then add their share of the total effects of the explosion event.

6.3 HUMAN BEHAVIOR IN FIRES

Fires and explosions are by their very nature traumatic events. The appearance of dense and irritating smoke and flames are the best indicators of impending peril and an obvious motivation for evacuation in case of fires (Tong & Canter 1985). However, even in buildings that are correctly designed and equipped with all the necessary exit facilities, some people are likely to become injured or killed. Fear of death, anxiety, disorientation, confusion and panic are just some of the direct effects of stress.

Under these circumstances most people (*ca.* 85%) react reflexively and intuitively (Wardlaw 1983). On evacuation they tend to use the same entrance corridors and the same everyday routine, even in cases where emergency exits are clearly marked. On the other hand, a smaller group (10–25%) is made up of people who show considerable self-control. They are able to perceive the danger and are able to work out and implement a realistic rescue plan. The third, equally small group. reacts irrationally and hysterically and are frequently the first victims of fires. However, panic behavior is the exception rather than the rule. Most people demonstrate unselfish behavior in trying to help others in fires (Bryan 1986).

An analysis of the behavior of people in fires in hotels, hospitals and nursing homes revealed that portable fire-extinguishers, although easily accessible, were actually not used (Wardlaw 1983). Out of 136 people evacuated, more than one hundred did not even try to extinguish the fire, while a few of those who did actually used the wrong extinguishing medium. Such behavior seems to be rather general, as results of a comprehensive British study confirm (Ramachandran 1985). The behavioral reaction of the mass determines to a considerable extent the behavior of the individual.

Thus it is clear that the behavior of people in fires is a complex process. More disturbing, in the case of fire, is that some people do not pay attention to fire alarms, either assuming a false alarm or assuming just another emergency exercise. Others overreact and panic, and try to escape using the first arrangement or solution available. Yet another group returns to the building under fire from which they had just been safely evacuated, thus endangering their lives again (Charbonnier 1984). In such situations, the role of the leader is crucial. He should be able to keep uninterrupted attention on all people, and to ask for and get their cooperation, enhance their sense of responsibility, ultimately leading them to a safe exit.

When confronted with a fire threat an individual makes a decision which is arrived at after several steps in the decision-making process. The steps include recognition, validation, definition, evaluation, commitment and reassessment (Bryan 1986). All of these steps are dynamic and stressful, with extreme stress experienced in the phases of commitment and reassessment. Two other conceptual models of the decision processes of the individual that have been developed in essence are similar to the structure of the decision-making process just described.

There is a significant difference in the behavior of males and females in first actions taken in fire (Bryan 1986). Male participants tend to attempt fire fighting, while females tend to engage themselves in warning and evacuation activities.

Analyses of the behavior of people in hotel fires have also revealed the formation of convergence clusters of hotel occupants (Bryan 1986), in particular rooms believed to be fire-safe shelters.

However, in view of the findings that the majority of fire victims in public assembly buildings, such as hotels and hospitals, are found dead in corridors, elevators, lobbies and stairways while attempting evacuation, a new, completely different safety philosophy is being advocated (*Fire Prev*. 1986). It suggests that in cases of fire in compartmented buildings of fire-resisting construction, people would be wiser to stay in their own rooms or units, awaiting a safe rescue, than to attempt evacuation by themselves. Although based only on one American study, the idea has aroused enough interest for further studies in other countries (*Fire Prev*. 1986).

7

Technical and technological protection measures

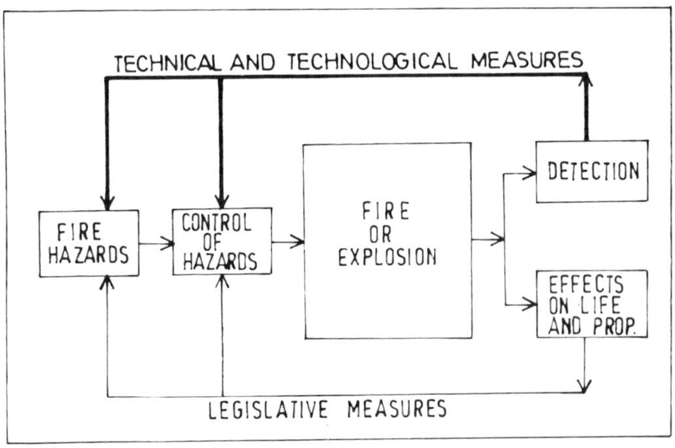

The role and significance of various technical and technologcial measures for both preventive and protective control of fires and explosions are evident from their position in the diagram depicting the system approach. In the preventive mode they ensure that no fire or explosion can be initiated from the simultaneous presence of fuel, oxidant and ignition sources(s), while in the protection mode they act upon the sources of fire and/or explosion hazards and their control in such as way as to extinguish the fire and explosion in the earliest stage, before any significant harm is done to people and property.

Fire and explosion protection measures can be systematized in many ways. One of the possible classifications can be made on the basis of the same chemical and physical fundamentals underlying combustion chemical reactions, as has been done in preceding chapters.

7.1 PROTECTION BY CONTROL OF FUEL

The presence of combustible material is one of the primary requirements for a fire or explosion. Therefore this reactant in equation (3.5) may be the one by which both preventive and protective control of fire or explosion can be accomplished.

There are several kinds of technical and technological measures to control fuel as a fire and/or explosion hazard.

(a) The simplest measure of preventive fuel control could be achieved if all flammable and/or combustible materials were absent. In practice, however, it is highly unlikely that an entirely fuelless environment and surrounding is possible, since at least some of the materials or elements in the open or in an enclosure are combustible and can support fire.

(b) A version of the above measure would be the substitution of combustible materials for noncombustible ones, or with materials possessing more acceptable fire and explosion characteristics. such as retarded combustibility, slower speed of flame propagation, lesser amounts of smoke, higher oxygen index, lower volatility, etc.

(c) A related measure is the physical removal of combustible materials in case of fire. In many instances if fire had not spread and reached the fully developed stage the usual practice of fire protection includes the removal of as yet unignited fuel from the fire scene. For example, in fires occurring in warehouses, or some other places where valuable commodities are produced, handled or stored, the items which have not been ignited are taken away before fire reaches them. The same principle of fuel removal may be realized in fires of liquids and gases, where the stopping of fuel from a container or a pipe is frequently the only practical means of fire control.

(d) Preventive fuel control can also be realized by making the fuel less combustible by treating it with a fire retardant. In this way the ignition temperature of the fuel can be raised, or the fuel can be made incapable of burning in its normal burning mode (flaming or nonflaming). Fire retardants are various inorganic chemicals which can make a flammable and/or combustible material less susceptible to ignition and fire spread. They are added in small quantities and usually act as chemical inhibitors. The inhibition can be explained by one or more of the following mechanisms (Shaw 1986): thermal (reduction of thermal buildup of the treated combustible), coating (formation of an insulating cover over the surface, which prevents oxygen reaching the fuel surface, gas formation (decomposition of chemicals with the release of nonflammable gases such as water vapor and carbon dioxide) and chemical (reduction in formation of flammable gases and liquids during the burning). The treatment can be performed by chemical changes on a molecular level (used with plastics and some synthetic fibres), by impregnation in the solution of dissolved or dispersed chemical(s), or under pressure (used with paper, wood and building materials), and by coating (used with nonabsorbent building materials).

Although no fire retardant can make a combustible material completely noncombustible the practice of treating combustible materials with a fire retardant as a means of fuel control is frequently exercised. Fuel control by fire retardants is obviously possible only with ordinary solid materials, such as textiles, plastics, wood, paper, etc. Fire retardants are also used in fighting forest fires, by pouring a retardant solution over the trees in the vicinity of the fire, thus slowing its advancement.

An indication of the resistance to burning of solids treated with fire retardants can be obtained by conducting laboratory tests on the rate of flame spread and by using the oxygen index.

(e) Another preventive method of fuel control is the reduction in quantity of combustible material to be handled, transported or stored. This is usually regulated by codes and standards and includes such measures as the use of minimal quantities of material required, dikes around fuel storage tanks, drains, etc. Fuel control by control of the fire load in buildings, stores, warehouses, industrial plants, etc., is a good preventive measure.

The prevention of formation of an ignitible mixture of a vapor and air can be accomplished by the reduction of the vapor pressure of the combustible. The reduction is brought about by cooling the combustible to a temperature below the flash point.

(f) Fuel dilution and removal is one of the proven methods of both preventive and protective fire and explosion control. For example, if a combustible liquid, soluble in water, is set on fire the simplest means of fire extinguishment can be accomplished by liquid dilution with water. The dilution of a flammable liquid leads to a reduction in the vapor pressure of the liquid to the point that a sustained burning can no longer be supported.

The same principle can be employed for the prevention of fires and explosions of flammable gases, vapors, dusts and mists within an enclosure by ventilation. Ventilation, defined as the process of supplying or removing air by natural or mechanical means, to or from an enclosure, may be general and local. General ventilation dilutes and removes combustibles through the entire volume of the enclosure, wheras local ventilation removes combustibles from a localized portion of space in the immediate vicinity of combustible emission. Either way, properly designed, operated and maintained ventilation keeps fuel concentration well below the LEL (lower explosive limit) values. Local ventilation is usually more effective than general ventilation. Codes and standards which regulate production, handling, transporation and storing of flammable and combustible liquids, gases, vapors and dusts specify the number of effective air changes within the enclosed space, usually with five air changes being the lowest required number. This number appears to be optimal regardless of how fuel concentration in the enclosure has been built up, i.e. either as a result of an accidental, massive release, or as a result of a continuous release. In the former case, fuel concentration, C, after some time following the release, is related to the initial concentration, C_o, and the number of effective air changes, KN, through equation (7.1) (NFPA 69 1978):

$$C = C_o e^{-KN} \tag{7.1}$$

K is a mixing efficiency factor (ranging from 0.2 to 0.9, depending on ventilation arrangements), and N is the number of theoretical air changes. The graphical form of equation (7.1) is shown in Fig. 7.1.

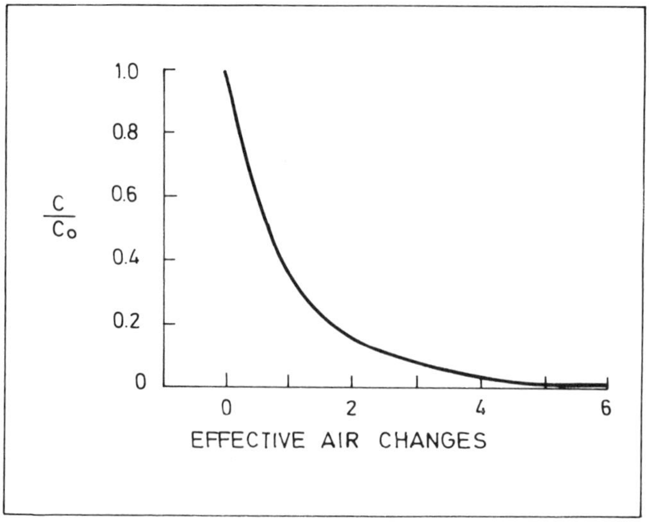

Fig. 7.1 — Reduction in fuel concentration as a function of effective air changes.

In the latter case (that is, the continuous release of fuel) the initial concentration rapidly increases with the increase in air change rate, and after about five effective air changes reaches an approximately constant value, dependent only on the volumetric flow rate of fresh air, and not on the air change rate. The mathematical expression for general ventilation in case of the continuous release is given by equation (7.2), and is shown in Fig. 7.2:

$$C = \frac{G}{Q}(1 - e^{-KN}) \tag{7.2}$$

G is the volumetric flow rate of fuel, L/min., and Q is the volumetric flow rate of air, L/min.

(g) Removal of fuel can also be accomplished by displacing flammable gases, vapors, mists or dust, in a tank, or container, by water, air, or an inert gas (so-called 'purging'). Streams of escaping gaseous fuel can be directed, diluted and dispersed by a carrier fluid (for example, water in the form of steam, spray, or mist) (Keith 1985).

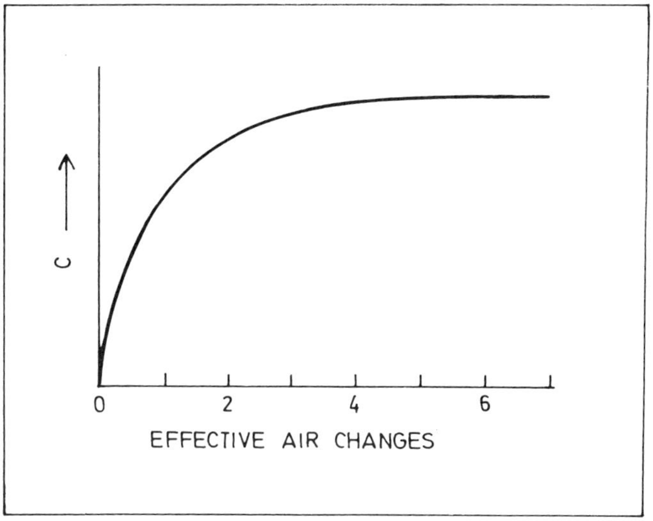

Fig. 7.2 — Fuel concentration as a function of effective air changes in case of continuous release of fuel.

(h) Prevention of fuel escape. Liquid and gaseous fuels should at all times be kept in closed and sealed containers, capable of preventing the escape of vapors and liquids even if upset. Containers should preferably be stored outside, in the open. If they are stored inside, they should be properly separated from the rest of the building area by fire-resistive construction elements (walls, floors, ceilings) and located at grade level. Smaller amounts of combustible liquids should be kept in specially constructed safety cans, designed to minimize the probability of accidental liquid or vapor spillage.

(i) Means of fuel arrangement. This measure of fuel control is mainly applicable to solid fuels, since the behavior of solid fuels in fire is detemined to a significant degree by the fuel array in the storage. Horizontal, vertical, cribbed or piled arrays all present different fire hazards.

(j) Good housekeeping, as reflected in tidiness, order and proper waste control, reduces combustible load and continuity and thus decreases the fire and explosion hazard.

(k) Proper, regular maintenance and service of equipment guard against accidents which release large amounts of combustibles.

(l) Tests for the presence of flammable gases, vapors, and mists by corresponding detectors provide early warning for the accumulation of dangerous concentrations of combustibles.

7.2 PROTECTION BY CONTROL OF OXIDANT

The control of oxidant concentration is both a preventive and a protective measure whereby the concentration of oxygen in air in confined, or nearly confined, vessels, equipment and rooms containing flammable gases, vapors, mists or dusts is reduced to levels which render fuel/air mixtures noncombustible in practically all cases. The reduction can be effected by purging the enclosure with a gas or a gas mixture deficient in oxygen. This measure is used almost only in industry where processing equipment (for example, chemical reactors, driers, mills, ovens, mixers, cyclones), storage equipment (tanks, vessels) and materials handling equipment (pneumatic conveyors, pipelines, pumps, etc.) are to be protected. Distribution patterns of the gas used for purging must be such that all parts of the enclosure being protected must have the minimum concentration designed for fire and explosion protection. Of course, complete rooms and buildings may be protected by this technique. If the oxidant is not in gaseous form, but is chemically bound as in oxidizing chemicals, purging the enclosure with an inert gas is not applicable.

There is a number of gases which may be used for purging and inerting. Among them the most widely used are carbon dioxide and nitrogen since they are relatively cheap, readily available and can be easily applied. Helium and argon are sometimes employed, but are more expensive and are applied only when a complete chemical inertness of the resulting atmosphere is required. Other gases, or mixtures of gases, which can be used for oxidant reduction are steam, halogenated hydrocarbons (halons) and stack or flue gases.

Inert gas can be applied either in a batch or in a fixed volume or continuously. The first method is occasionally employed, mostly for inerting containers filled with a flammable gas, while the second method is employed when a safe atmosphere is required during normal operational cycles. In the continuous method, purging gas may be applied either at a fixed or at a variable rate, depending on the gas demand. Regardless of the method of application employed, safety of personnel should be ensured, for exposure to oxygen-deficient atmosphere, or to toxic products in a gas mixture, may pose a serious threat to life safety.

For purging to be effective, the normal oxygen concentration in air (21%) should be reduced to below a specific limit. This limit is variable, depending on the character of the combustible and of the purging, or inerting, gas. For most hydrocarbons, ignition is prevented when the oxygen concentration in nitrogen/air mixtures is reduced to 10–12%, or to 14–15% in carbon dioxide/air mixtures. The recommended oxygen percentage are normally 2–3% lower as a factor of safety. However, it is best to maintain oxygen concentration at an as-low concentration as practicable. The effectiveness of purging or inerting is checked by measuring the remaining oxygen concentration with an oxygen analyser, or by analyzing the quality of the purging or inerting gas.

The principle of oxidant reduction by an inert gas can be successfully employed in fire extinguishment. The gases used, mainly carbon dioxide and nitrogen, help to extinguish fire in part by other extinguishing mechanisms, such as fuel cooling, but the principal mode of action is the reduction of the oxygen content of the atmosphere by dilution. The high effectiveness of common inert gases is explained also by their relatively high thermal capacities which make them good thermal absorbents (Drysdale 1985).

Nitrogen gas is used for fire extinguishment only through fixed entinguishing systems, but not so widely as carbon dioxide or halons. On the other hand, carbon dioxide may be used either in portable fire extinguishers, or in fixed systems.

Portable CO_2 extinguishers (NFPA 10 1984) can be used for the manual combat of small incipient fires. For some time they were the leading fire extinguishers, until the development of equally or more successful portable extiguishers based on dry chemical agents and halons. However, carbon dioxide still finds broad application in fixed extinguishing systems used for extinguishing fires of flammable gases and liquids, electrically energized equipment, vehicle engines and, in some cases, of ordinary combustible solids. The method of CO_2 application depends on the environment which is being protected (NFPA 12 1985). In completely enclosed spaces, such as rooms, containers, ovens, machines, etc., which enable the required CO_2 concentration (between 30 and 80% by volume, depending on the combustible) to be built up and maintained until the fire has been completely extinguished, total flooding systems are used. They can extinguish not only surface fires but also deep-seated fires. Fires in semienclosed and open spaces can be extinguished by local application of large carbon dioxide doses directly on the surface of the burning material. Details on designing, installing, testing, operating and maintaining carbon dioxide fixed extinguishing systems can be found in corresponding fire protection standards.

7.3 PROTECTION BY CONTROL OF IGNITION SOURCES

Control of sources of electrical heat energy

Control of Coulomb heating and arcing. The principles of control of heating and arcing as evident signs of electrical faults are fairly simple. One of the basic requirements is that all electrical installations should be made, used and maintained in conformity with existing electrical standards. For example, electrical circuits should be adequate in number and capacity and should be clearly identified. Conductors should be of the right size and covered by the right type of insulation. Old and deteriorated electrical conductors have to be replaced since the maximum current they carry rapidly decreases with aging of the conductor. Only fuses of the right size for the circuit they protect should be used.

A common mistake appears in the overloading of electrical circuits by placing too many appliances on a single circuit. Another poor practice is to let unqualified people do wiring modifications instead of qualified technicians or servicemen.

A good preventive measure against electrically caused fires is to avoid the use of extension cords. Their use can be minimized if a sufficient number of electrical outlets are installed at convenient locations.

Electrical apparatus and appliances that use electric condutors as heating elements or that generate heat by electrical arcing must be properly installed and maintained. This means, among other things, that ample air circulation is necessary to prevent the development of hazardous temperatures. At no time should such devices be covered or blocked. Although it is assumed that the cleanliness and regular maintenance of electrical apparatus and appliances should be kept in due condition, in practice these are often overlooked.

Electrical codes specify the type of equipment to be installed in a location. In hazardous locations only special electrical equipment is allowed (NFPA 70 1984). A good practice is to have circuit diagrams for electrical appliances and equipment readily available for maintenance purposes. More sophistiated equipment should be operated strictly according to corresponding manuals.

Control of heat from lightning. Lightning discharges cannot be prevented. The control of this source of ignition is possible only through a lightning protection system. Although there may be numerous variations in the system details, the main function of the system — to provide a means by which a lightning discharge is attracted and safely conducted to earth before it can strike the structure protected — should be fulfilled in all of them. Atmospheric electricity discharge on touching the uppermost parts of a structure travels throughout the structure to earth using a path of lowest impedance. Since metals have much lower impendances than most building materials, such as stone, wood, concrete, brick, etc., common practice is to install a lightning protection system made of metal. The main components of such a system are air terminals, and ground terminals, buried in the ground. These two terminals are connected by a conductor system (NFPA 78 1983) (Fig. 7.3).

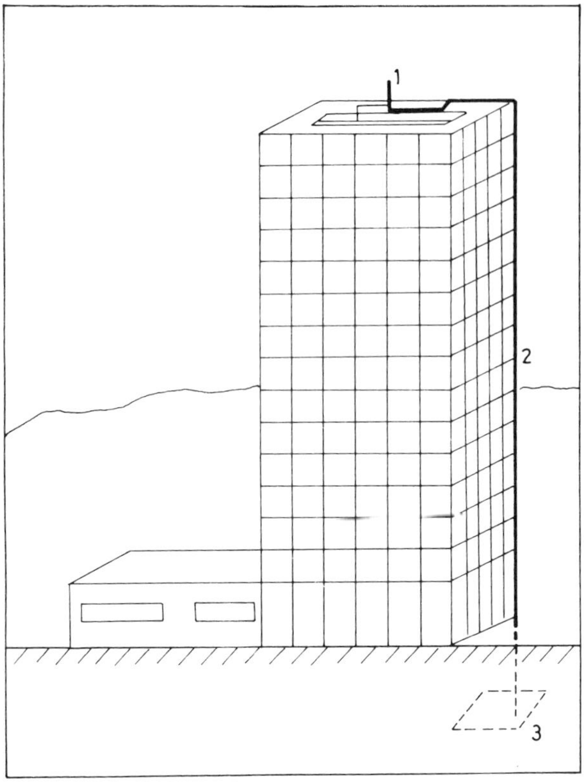

Fig. 7.3 — Principle of lightning protection of a building; 1. air terminal, 2. bonding conductor, 3. ground terminal.

Air terminals should be placed high enough above the structure to avoid ignition of the combustibles by electric arcing. It is best if the conductors provide the most direct path between air and ground terminals. If this is not possible, they must not bend, or loop, sharply.

Ground terminals should secure good contact with the soil. Actual arrangements, whether in the form of a rod or an elaborate wire network, depend on soil conductivity. In moist, good conducting soils, such as clay, a single rod driven into the ground may be satisfactory. In poorly conductive soils or bedrock, more elaborate means (for example, a buried wire grid, or ring conductors) are usually required. Grounding may be accomplished simply by connecting conductors of the protection system to buried metallic water pipes.

Control of static electricity. Where a hazard of accumulation and spark discharge of static electricity exists, modes should be provided to protect combustibles against possible ignition. Static electricity can be dissipated or removed by several means. The essence of protective control is in supplying natural or artificial conducting paths which prevent the accumulation of static electricity, or if this has accumulated, to enable its safe dissipation or removal. This can be done (NFPA 77 1983) by:

— bounding and grounding
— humidification of the atmosphere
— ionization of the air

Bounding and grounding of conductive objects prevents accumulation of static electricity. Potential differences between individual conductive objects are significantly minimized by bonding them, while grounding decreases potential differences between objects and the ground.

The conductivity of some ordinary materials which are poorly conductive and therefore prone to static electricity accumulation (for example, concrete, masonry, fabrics, wood, paper, etc.) is to a large extent controlled by the relative (not absolute) humidity of the surrounding air. Such materials reach equilibrium with moisture relatively quickly. At relative humidity values higher than 50% enough moisture is accumulated on the surface to make the material a much better conductor. Thus if the surrounding atmosphere is humidified to a relative humidity value of at least 50%, a more dangerous accumulation of static electricity may be prevented. However, at higher temperatures, the relative humidity decreases considerably, which means that humidification cannot then be effectively used as a protective measure. Nor can it be used if the surface of the material which is being protected is mechanically separated from the surrounding air (for example, by coating it with a thin film of organic material, such as grease or oil). Humidification may not be feasible in all kinds of structures in which static electricity may be a hazard source. Furthermore, it cannot be applied in those cases where materials are sensitive to high water content in air. Therefore, humidification, although a natural and elegant means of static electricity dissipation, cannot be a universal cure for static electricity problems.

Static sparks jump between electrified objects because the air in which they are immersed is a very poor conductor. However, if air in the vicinity of charged objects is ionized in some way then conductivity of the air may be increased to a point where it can effectively drain the accumulated static electricity. There are several devices and means by which static electricity can be 'neutralized'. The most frequently employed are a 'static comb" (a metal bar which contains on its surface a number of sharp-pointed needles) and an electrical neutralizer (a high-voltage device). They are capable of ionizing air in the vicinity of the protected body, thus removing accumulated static electricity. In some specialized cases, ionization of the air can be performed by radiation from a radioactive source, or by employing an open flame. While these means can be fairly effective they are not widely used because of environmental and fire protection objections.

The hazard of static electricity accumulation in highly hazardous places can also be minimized by employing static conductive floors and requiring workers to wear conductive shoes or clothes.

Control of mechanical heat energy

Heat from the friction which results from old and worn parts of equipment can be prevented by replacing the old parts with new parts. Also important are the regular inspection of the equipment and the removal of combustibles, such as dirt and grease, from the site.

The formation of sparks can be prevented by using nonferrous, nonsparking metals, or nonmagnetic nonsparking stainless steel (Powell 1986). Metals with a negligible spark hazard include some alloys, such as bronze and copper–beryllium, nickel and aluminium. Therefore, when there is a hazard of spark formation in hazardous locations, tools made of these materials should be employed. If it is possible to use plastic or wooden tools then the danger of sparks no longer exists.

Sometimes it is not possible to avoid the danger of spark formation and its entering into the dust or vapor collecting system. In such cases, ferrous particles can be removed at the entry point by strong, permanent magnetic separators. The hazard can be further reduced by employing conveyors or ducts fabricated from nonferrous nonsparking materials. Nonsparking materials should also be used for the construction of blades and the spider of the fan and the fan housing.

Control of chemical heat energy sources

Various chemical heat energy sources, such as open flames, smoking, hot works (welding and cutting), direct-fired equipment, etc., are closely associated with a wide range of human activities. Althought it is obvious that some of these ignition sources can be (and are) easily regulated the fact is that a large percentage of all fires and explosions are attributed to them through human carelessness and negligence. For example, although it is more than obvious that open flames from matches, lighters, cigarettes and some space heating devices are a source of fire and explosion hazards, in many locations people still bring them into contact with combustibles. This is most often because of gross negligence, but sometimes on purpose, as in cases of incendiary ignitions and arson.

Most of the measures to control chemical heat energy sources should be self-evident. Thus, direct-fired equipment, heated surfaces and open flames should not be allowed in enclosures containing flammable gases, vapors, mists or dusts. Such equipment should have built-in flame arrestors and flashback protectors. If unnecessary equipment and processes develop temperatures high enough to ignite flammable and combustible materials, then these should be banned.

Welding and cutting operations should be performed strictly according to corresponding fire standards. This means that only listed equipment in good working condition is recommended and that special attention should be directed to the protection of the work area.

7.4 PROTECTION BY CONTROL OF COMBUSTION PRODUCTS

7.4.1 Control of heat

Heat is one of the principal combustion products. It is transferred from the fire scene by convection, radiation and conduction and is responsible for the spread of fire. However, the transfer of heat can also be the basis for fire extinguishment. The objective is to remove heat, often just a small fraction of the total, from combustibles to the point that flammable vapors and gases are no longer liberated and the fire is thus extinguished. This can be accomplished either by the cooling of combustibles, which are either burning or hot, from exposure, or by isolating the fire using fire-resistive barriers, or by spatial separation of the fire from nearby combustibles.

The *cooling action* can be realized by applying an agent which has good thermal characteristics, such as high specific heat capacity, high latent heat of vaporization and high boiling point. Of the few candidate liquids, water is superior. Although its specific heat capacity (4.186 kJ/kg K) and latent heat of vaporization (2254.8 kJ/kg) are not the highest, they are nevertheless high enough to make it a very efficient cooling agent (Hodnett 1986a). The indisputable advantage of water is that in most cases it is readily available. Historically, water has long been and still is the extinguishing agent most used.

In addition to cooling effects, which are the primary mechanism for stopping fire, water extinguishes fires also by diluting the oxygen concentration in enclosed spaces. The dilution effect is a consequence of steam formation on the contact of water with hot materials. The high steam/liquid ratio (*ca.* 2500:1) suggests a high diluting capacity of steam. However, this effect is transistory, as the steam, being of a low density, quickly diffuses from the fire area.

For water to be an effective extinguishing agent it has to be applied directly to the surface of the burning fuel. The degree of cooling is strongly dependent on the way the water is used. Solid stream is less effective than droplets, and less effective still than spray. The best effects are obtained with small spherical droplets of uniform size (0.3–1 mm). Droplets of smaller size, although the best coolant are difficult to apply because of increasing air resistance.

For most purposes, plain water is satisfactory. However, some of the properties of water, such as its relatively high freezing point, high surface tension, low viscosity and electrical conductivity, may reduce its effectiveness as an extinguishing agent in some situations (Rosander & Giselsson 1984). For example, if the ambiental temperatures are around 273 K or lower, water freezes and becomes useless. Problems of freezing can be eliminated by a number of means, including the use of dry, instead of wet, pipe systems, the heating and the addition of some chemicals which decrease the freezing point (Hodnett 1986a).

The high surface tension of water which reduces its ability to thoroughly wet and penetrate burning combustibles because of high run-off, and which lowers the total free surface area available for cooling action, can be reduced by the addition of some chemicals. The chemicals should have good properties with regard to corrosiveness, toxicity and stability when mixed with water.

In extinguishing some types of fires (for example, forest fires) it is desirable that water be able to stay in contact with the combustibles for as long as possible. This can be effected by making water more viscous ('thickened'), again by the addition of some chemicals. The distance a water stream reaches from a fire hose can be considerably increased by using water to which some organic polymers have been added. They significantly reduce the problem of friction of water molecules and thus increase the nozzle pressure.

The electrical conductivity of water precludes its use for extinguishing fires involving live electrical installations and equipment of high voltage. In such cases it is best to avoid the use of water and to use an alternative extinguishing agent. If water is nevertheless used it should be applied from a safe distance.

Although water is the most universal of all extinguishing agents there are some materials which cannot, or must not, be extinguished with water. These are mostly certain chemicals, some metals and radioactive materials (see section 2.1). In fires of gases and combustible and flammable liquids, water is mostly used for the control of heat transfer to nearby combustible materials and structures.

Water for extinguishing purposes should be available from a reliable water supply system in an amount which satisfies both normal urban and industrial demands and fire protection requirements (Anderson 1986a). In most cases it is necessary for water in the system to be kept under pressure (usually 450–550 kPa). The reliability of the supply system is ensured by duplicating vital system units, such as pumps and storage facilities, and by regularly conducting a well designed maintenance and testing program.

Systems for water distribution can be separate (only for fire protection) or combined (dual), that is for both normal consumption demands and for fire protection (Anderson 1986b). As sources of supply, either ground water or surface water can be used. Water is transported from supply sources to consumers by a distribution system. The pressure necessary for the distribution can be developed either by gravitational force or by pumps, or by a combination of these.

The simplest way of applying water to a fire is manually, by hose lines either from outside hydrants or from standpipe and hose systems lcoated inside the building. While these systems have been useful and are still found in many types of structures they are gradually being replaced by automatic sprinkler systems.

Automatic sprinkler systems are currently being considered as the most effective means of extinguishing or controlling fires in hazardous occupancies (Hodnett 1986b). When designing sprinkler protection systems the nature of the hazard and building features should be considered.

The advantages of automatic sprinkler protection over manual protection are numerous. Many reports and fire statistics unquestionably prove that sprinklers are especially effective in life safety and in the protection of property (Ruegg & Fuller 1987). Yet the application of automatic sprinklers in Europe is lagging behind in comparison with the USA and Canada. A comparison for 1983 revealed that the total number of sprinklers installed in Europe was *ca*. 2.2 million vs. *ca*. 14 million in the USA (Young 1985).

An automatic sprinkler system consists of a system of pipes fed by water from a reliable source and a number of sprinklers placed at intervals along the pipes. Water is discharged automatically from sprinklers in the form of fine droplets. The automation of sprinkler reaction is achieved by employing a thermosensitive device (essentially a heat detector) as an integral component of the sprinkler. When the temperature of the detector reaches a predetermined value, water is automatially discharged onto the fire. The working principles of an automatic sprinkler system are shown in Fig. 7.4.

Depending on whether the pipes contain water or air under pressure a sprinkler system may be designated as a wet pipe system or a dry pipe system. Dry pipe sprinkler systems are used in locations where there is danger of the water in the pipes freezing. Alternatively, wet pipe sprinkler systems with antifreeze solutions in the pipes can be used.

Ordinary sprinkler systems are designed such that heat from the fire activates just one or a few sprinklers. Indeed it has been found that in more than 50% of extinguished fires in sprinklered occupancies, fewer than five sprinklers have been activated (Young 1985). However, there are sprinkler systems, termed deluges, which have all sprinklers open at all times, thus deluging the entire protected areas. They are used in enclosures containing very high fire hazards (for example, in storage of flammable liquids). Sprinklers are widely used in industry and high buildings. In some countries they are required even in new residential buildings. However, if not maintained properly and regularly the obvious benefits of sprinklers may be lost (Todd 1987).

Automatic sprinkler systems may be designed to be an integral component of a fire alarm system (see section 5.3). When water flows through risers or mains supplying the system, an alarm notification is given.

Water spray systems are similar to automatic sprinkler systems. Water is discharged in the form of a mist, not through sprinklers, but through specially designed nozzles or devices (Hodnett 1986c). They are employed for the protection of flammable liquid and gas storage and transporting units and electrical transformers. If properly designed and maintained, water spray systems in some situations can be used as a substitute for fire-resistive elements in firewalls and floors.

Another means of heat control is by *structural fire prevention measures*. They must confine fire to the area of origin, thus protecting people and property not directly exposed to the fire. The degree of confinement depends on the intended purpose of the fire confinement and on the expected severity of the fire (Campbell 1986, Bamert 1985). If the main purpose is to preserve the structural integrity of a building then barriers must endure the maximum expected fire that would develop inside the enclosure. However, if the barrier is to resist the fire only until people or valuable goods are evacuated, then the requirements put on it are less demanding.

Fire-resistive barriers are tested in laboratory furnaces subjected to a time–temperature regime, known as the standard time–temperature curve, and assigned a fire-resistance rating, expressed in hours, or minutes. Unfortunately, test results are poorly reproducible even in the same laboratory, so the actual fire resistance may be far from the rated value (Vandevelde 1985). Examples have been given of some real fires in which a building with barriers having a fire-resistance rating of 30 min did not collapse, whereas in another case, a building, having fire barriers rated as 90 min, could not control the heat from the fire and disintegrated after 30 min (Bamert 1985).

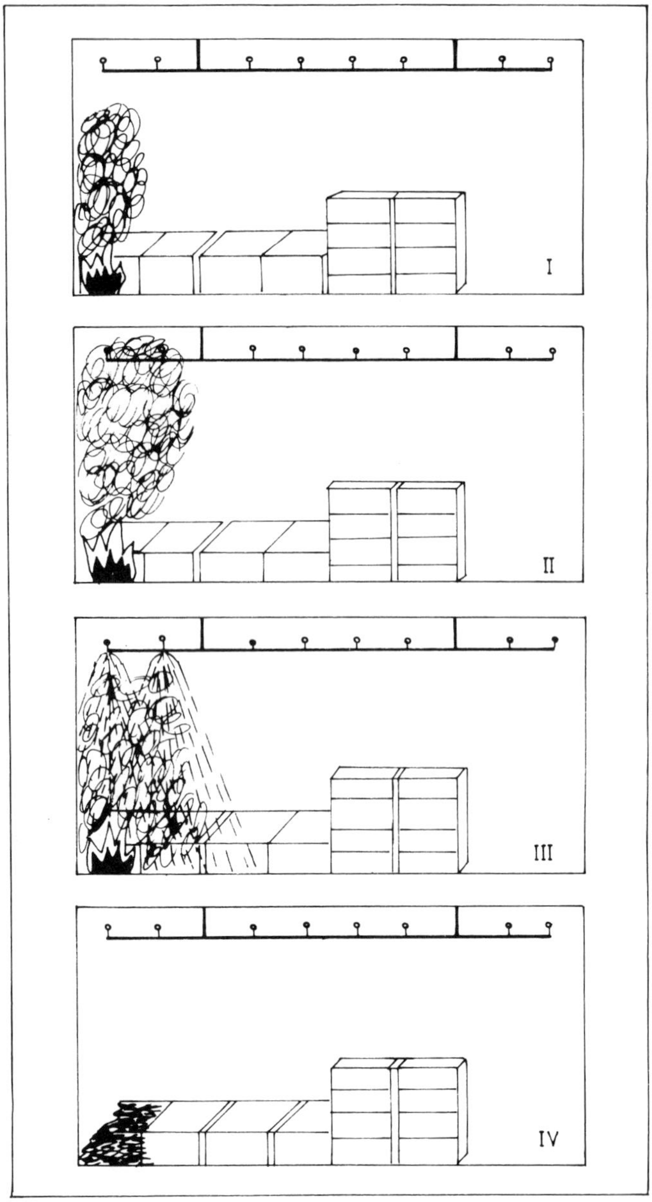

Fig. 7.4 — Principle of automatic sprinkler protection: I. fire develops, II. fire reaches sprinkler heads, III. sprinkler activates, IV. fire extinguished.

Because experimental testing is expensive in some countries, fire barrier ratings are allowed to be determined by analytical calculations (Kirby 1986) on account of relatively good agreement between calculated and experimentally measured fire resistance.

In many cases heat from fires, and fire spread, cannot be successfully controlled by building compartmentalization since the barrier is often left unprotected. An open door, unenclosed shaft or concealed space provide heat transfer to other parts of the building, nullifying the role and effects of the barrier.

Heat from fire can also be controlled by *spatial separation*. This principle is one of the means of exposure protection. Knowing the duration of the fire and the total heat released, the separation distance for neighbouring combustibles can be calculated (Law 1963). In practice, site conditions and economic constraints frequently make a structure difficult, or impossible, to comply with the required distance. In such cases there are numerous alternative means of controlling heat from the fire (NFPA 80A 1980.

7.4.2 Control of flame

Flaming combustion is characteristic of fires of flammable and combustible liquids and gases. Of course, many solid materials can and do burn with the appearance of flames, but usually in a later phase following the ignition.

There are two main methods by which flaming fires can be extinguished. One of them involves the exclusion of air as an oxidant in the fire atmosphere, and the other is the inhibition of chemical reactions occurring in the flames. The exclusion of air as an extinguishing method based on the use of gaseous agents has already been described in section 7.2. It can be applied to both flaming and nonflaming fires. Another effective means of exclusion of air can be realized by the use of various water solutions of some organic substances which, when aerated and agitated, produce copious amounts of bubbles in the form of foam. The foam, being lighter than the liquid phase, floats on the surface of flammable and combustible liquids, thus preventing both the supply of air to the fire and the vapor transmission from the liquid surface.

Depending on the nature of the foam concentrate and the method of generation the volume of foam may be many times greater than the volume of foam solution. The ratio "volume of foam/volume of original foam solution" may be made to vary from 1 to 1000. For practical purposes three subdivisions, (1–20):1, (21–200):1, and (201–1000), are usually designated and used. They correspond to low, medium, and high expansion foam, respectively. Each of these types of foams is used for a specific type of fire. For example, low expansion foam is used primarily to extinguish fires of burning liquids spills or reservoirs. The dense, coherent foam blanket, in addition to excluding air, cools the liquid surface and thus diminishes and prevents the formation of flammable vapors. The medium and high expansion foams, because of their large volume, are suitable for extinguishing fires in deep enclosures, such as basements vaults, pits, and inside areas of industrial equipment. Such spaces can be quickly filled with foam to the required depth, including total flooding, if necessary.

To make a foam, only a few percent of liquid concentrate is needed. Usually 1, 3 and 6 percent concentrations are made. As the solvent, both fresh and seawater can be used. The first foams were produced by using proteins from some natural products. Protein type (P) concentrates are still used, but many more agents, predominantly synthetic, with better characteristics have been developed. They include fluoroprotein (FP), film-forming fluoroprotein (FFFP), and aqueous film-forming foam (AFFF) agents as well as formulations designed for fighting fires at low temperatures and for fires of liquids soluble in water (so-called alcohol-type foaming agents). A combination of AFFF and alcohol-type concentrate, "light water", has been claimed to be the most versatile foam (*Fire J.* 1985).

In addition to the air excluding function, foams, owing to the water content, also extinguish fires by cooling the liquid surface and by displacing air with steam. However, the latter mechanisms are only of minor importance. The blanketing characteristic of foams can also be used for the prevention of evaporation of spilled hazardous liquids, such as chlorine, or ammonia (Waters 1985).

The extinguishment of fires by flame inhibition can also be accomplished if some chemicals, capable of entering into the chemical reactions making the flames, are added in sufficient concentrations. These chemicals can be divided into two basic groups: anorganic and organic. Anorganic-based chemicals come in the form of dry powders, while organic chemicals are mostly liquid. The precise mechanism(s) by which these materials extinguish fires are still unresolved and are the subject of continuing research. It is believed that the most probable is a chain-reaction-breaking theory (Haessler 1974). According to this theory, flames are a result of a complex series of chemical reactions occurring during the combustion and involving the presence of highly active free radicals. The radicals, being both products and reactants of reactions, act as "chain carriers". When some chemicals are introduced into the flames the sequence of chain reactions is abruptly interrupted and the flaming fire thus extinguished. It is found that the most effective chemicals are those containing either highly electropositive (alkali) or electronegative (halogenides) elements.

Sodium and potassium salts, such as bicarbonate, carbamate and halide, make a group of dry-chemical-extinguishing agents often referred to as "ordinary dry chemicals", or "ordinary dry chemical" powders. They can extinguish any flaming fire (Drinkwater 1985) but are usually intended for the extinguishment of fires involving live electrical equipment. However, in areas where sensitive electrical equipment (computers, telephone exchanges, control rooms) is located, dry chemicals should not be used since, because of insulating characteristics, they might render such equipment inoperative. Dry-chemical-extinguishing agents may be applied by any of the usual means of application, that is through fixed systems, hand hose-line systems and portable extinguishers. In all such cases a gas, known as an expellent, must be used to discharge and disperse the powder into the hazard area.

The electronegative chemical elements (halogenides) are used for fire extinguishment in the form of some organic compounds. The simplest are halogenated derivatives of methane and ethane, termed halones. Changes in fire and explosion characteristics of the parent compounds brought about by the introduction of just one atom of a halogen into the molecule are striking indeed. Thus, while both methane and ethane are flammable and explosive, methyl bromide is not only nonflammable but is a good extinguishing agent. To simplify the names of many possible combinations of methane and ethane with flourine, chlorine and bromine, a numerical nomenclature was proposed and has been in use since (Moore 1986). Carbon atoms in the molecule are numbered first, followed by the second, third, fourth and fifth digits, representing the number of fluorine, chlorine, bromine and iodine atoms, respectively. If a halogen is missing, a zero is assigned, except in the case of iodine when zero is not expressed. For example, dichloromethane is assigned the number 1020, methyl iodide 10001, bromotrifluoromethane 1301, etc. The first halogenated agents found to have extinguishing properties were carbon tetrachloride (halon 1040), methyl bromide (halon 1001) and chlorobromomethane (halon 1011). They were used for some time until it was realized that they can be highly toxic. The choice of a halon to be used for fire fighting is a compromise between characteristics such as toxicity, extinguishing efficiency, chemical and thermal stability, and volatility. An ideal halon would be nontoxic, have high extinguishing efficiency, good thermal and chemical stability, nonexistent or low corrosivity, and high volatility. Halons, the nearest to this, are 1301 (bromotrifluoromethane) and 1211 (bromochlorodifluoromethane).

Halons can be applied manually from portable extinguishers for local application, or from total flooding systems for the protection of enclosed spaces. Although most of them are volatile enough to be self-expelling, overpressurization with an inert gas, such as nitrogen, is necessary for quick action and good diffusibility.

7.4.3 Control of smoke and gases

Many of the measures described earlier in this chapter directly, or indirectly, contribute to the control of smoke and hot gases from a fire scene. For example, an obvious and simple way of smoke control would consist of avoiding all the materials which, when ignited, liberate excessive amounts of smoke. Also, many of the measures for the control of heat (for example, automatic sprinklers, compartmentation with free-resistive barriers) also control smoke and hot gases. Therefore, only a few of the additional measures, designed and used specifically for the control of smoke and hot gases in enclosures, will be briefly mentioned.

One of the recognized and often used means for smoke and gas control in enclosures is ventilation. It is widely used in fire fighting operations since it effectively removes smoke, gases and heat from enclosed spaces (Grimwood 1985). In this way the flow of gaseous products of combustion can be directed to a desired direction, the fire scene cleared and visibility improved, thus protecting people and fire fighters.

Smoke and gases can be removed by using various types of unit vents and curtain boards. While the benefits of ventilation in sprinklered buildings are still controversial (Watermann 1984) ventilation of single-storey structures by means of automatic vents is very effective. Any opening on the roof or on the walls (for example skylights and windows) will function, to a degree, as means of removal of smoke and gases. However, automatic vents, specifically designed for smoke and gas removal, are far superior. They open automatically whenever the smoke concentration or the gas temperature reaches a predetermined value. If for any reason the automatism of the opening fails, means of opening the vents manually should be provided. Levels of smoke and heat necessary for a prompt opening of vents can be increased by using curtain boards. Dimensioning and spacing of vents, and the location and depth of curtain boards, are specified in corresponding guides (NFPA 204M 1985).

In multistorey buildings the control of smoke and gases can be performed by a building's air conditioning and ventilating systems. In compartmented buildings, in the event of fire ventilation fans should be shut down and smoke dampers in the ductwork closed. This means of smoke control is termed passive. Another approach, called active, is to make use of a building's heating, ventilating and air conditioning systems which can remove smoke and gases in an orderly manner. The principle of the active approach is to pressurize emergency exit stairwells and elevator shafts and thus prevent smoke movement and migration from the fire area to the other parts of the building. Details of active smoke control can be found elsewhere (see, for example, the ASHRAE manual, ASHRAE 1983). In more complex building interiors the removal of smoke and gases can be predicted only with the aid of computers (see section 4.5).

7.4.4 Control of pressure

Control of the effects of high pressures from explosions can be obtained by one or more of the following means (Bartnecht 1981, Palmer 1973):

— spatial separation of explosion hazardous units, or processes to precalculated distances from potential receptors of explosion impact. This method is normally employed whenever there is a permanent explosion hazard, for example in explosive-producing plants.

— explosion containment. Excessive pressure buildup can be contained in an explosion-resistant production unit, thus preventing the escape of any of the products of explosive combination. The containment of pressure is not applicable to larger equipment or structures, since they cannot sustain the range of maximum explosion pressures normally encountered.

  The design and functioning of this means of pressure control is very simple since no movable or operational part is involved. Containment is made possible by the use of materials with good mechanical characteristics which must be able to endure the explosion energy. The control of pressure by containment is practical only with gas explosions since maximum pressures, P_{max}, are relatively low ($P_{max} = 8P_o$, where P_o is the starting pressure of the gas mixture). Pressures produced from explosions of dense explosives, or from thermal runaway or decomposition reactions cannot generally be controlled by containment.

— explosion quenching, or explosion suppression. If chemical reaction(s) leading to explosion can be interfered with in some way then an explosion could be stopped before it reaches full development. The suppression can be realized either by heat removal or chemically, by adding a reactive material. The heat of explosion reaction can be removed by materials having a high heat sinking capacity and large surface area. These may be the walls of the equipment or some materials introduced into the reaction vessel. Another way to remove heat is to cool the gas mixture below the ignition point, or to dilute the gas mixture below the lower explosive limit.

  Explosion suppression is possible only if coupled with systems for explosion detection and injection of the quenchant. The control of pressure by explosion suppression is generally limited to homogeneous gas mixtures in which the suppressant can be uniformly and quickly dispersed. Explosions of dense explosives cannot normally be suppressed because the process of introduction and mixing of the suppressant is too slow compared with the rate of pressure rise.

— explosion dumping. Explosions can also be interrupted by a fast transfer of the reaction mixture to an alternative vessel, or into a medium which stops the propagating reaction. This method of pressure control is mostly applicable to dense explosives.

— explosion venting. In this method the explosion is not suppressed but a means of pressure relief is provided by a structural element (such as a blowout wall, rupture panel, burst disc) built into the sides or ceiling of the enclosure. The vent system should have a low inertia, that is be able to release gas from explosure approximately at the same rate at which the pressure is increased. When designing explosion venting systems, precautions should be taken to protect the outside of the vented location since the vented contents may pose environmental or life safety risks. Of course, explosion vents for buildings or plant units must operate at pressures significantly lower than those at which the building construction would collapse.

In practice, explosion venting is the usual means of pressure control. It is especially convenient for homogeneous rarefied explosions. Venting of explosions of heterogeneous mixtures (for example a liquid plus its vapor) is more complicated, as the discharge often involves both phases. Venting cannot be applied to the control of detonations.

The ratio of the vent area to the volume of enclosure protected can be found from experimental data obtained either by the Hartman bomb or by the 20 L sphere. The existing data have been used for the construction of numerous monographs from which the size of the event can be determined. For enclosures of relatively small volume (up to 100 L) a cube-root law holds (Field 1983):

$$\left(\frac{\mathrm{d}P}{\mathrm{d}t}\right)_{\mathrm{max}} V^{1/3} = K_{\mathrm{st}} \qquad (7.3)$$

where K_{st} is a constant, V is the volume of the enclosure and $(\mathrm{d}P/\mathrm{d}t)_{\mathrm{max}}$ is the maximum rate of pressure rise. The vent area is then determined from the corresponding nomograph.

Vent ratios for larger enclosures, such as buildings and industrial plants, can be calculated by a number of formulae. Anthony (1977/78) has examined several different approaches and concluded that the most promising is a mathematical model by Yao (Yao 1973). Yao calculated the vent ratios theoretically, on the basis of the law of conservation of mass and the assumption of adiabatic compression of the gas mixture. A simple equation useful for practical application is that given by Runes (Runes 1972):

$$A_{\mathrm{v}} = (CL_1 L_2)/P^{1/2} \qquad \mathrm{m}^2 \qquad (7.4)$$

where A_{v} is the vent area, C is a constant for the given fuel — ranging from 6.8 for propane, organic dusts and organic mists, to 17 for hydrogen — L_1 and L_2 are the smallest and next smallest enclosure dimensions respectively (in m), and P is the maximum overpressure which can be withstood by the weakest structural element (in kPa). The Runes equation holds under the condition that the largest enclosure dimension, L_3, satisfies the relation $L_3 \leqslant 3(L_1L_2)^{1/2}$.

7.5 MEASURES FOR LIFE SAFETY

All of the technical and technological measures in this chapter are implicitly for the function of life safety and property protection as the ultimate aim. However, some measures, explicitly concerning life safety, will be outlined here.

People spend most of their time in various enclosures, such as apartments, office buildings, shops, public places, etc. It is these places where the threat to life safety is the greatest. Therefore, good life safety protection should start with the design stage of buildings, taking into account numerous variables that affect the safety of the occupants. Of these, especially important are the interior layout, the type and fire characteristics of the finished materials, and building services and utilities (Fitzgerald 1986, NFPA 101 1985).

The architectural design of the interior layout of a building includes such considerations as means of primary and alternative egress, compartmentation of interior areas to fire-resistive zones and the protection of vertical openings and other places in the building such as stairways, elevator shafts and atriums through which the fire may spread. Exits from the building should be sufficient in number and protected against fire, smoke and heat during the time needed for evacuation. They should be of adequate capacity and designed on the maximum probable population of the building. Travel distances to exits should not be excessive. It is especially important that exit access should be as short and direct as possible and be free or unobstructed at all times. A common mistake, often leading to a tragic outcome, is to lock the exit doors while the building is occupied. Exit doors must always be openable from within the building. Clear exit signs, visible from all directions, should be posted along the paths to exits and on exits themselves. Means of egress should be continuously illuminated and furnished with emergency lighting.

If primary exits become obstructed or blocked by products of combustion, alternative means of egress should be provided. In this respect it is good practice to provide the building with large openable windows, as they provide a number of benefits in the fire.

In designing life safety measures for buildings, not only the nature and dynamics of possible fire in buildings should be taken into consideration, but also the characteristics of the occupants, such as their density, mobility, the degree of awareness, age, and expected behavior in emergencies. In all cases when the occupants are alert and mobile, evacuation from the building is the most common means of life protection. However, when the building is of such size and design that successful evacuation of all occupants would be unlikely it is better to try other alternatives. One of these might be protecting the occupants in place until the arrival of rescue units. This approach may be applicable for buildings whose occupants are restricted in movement, such as hospitals, prisons and day-care centers. An analysis of 14 large fires in buildings clearly showed the benefits of nonevacuation is such cases (Macdonald 1985). The other alternative might be to provide a safe area for temporary refuge in a section of the building. This is, in fact, a combination of the former two approaches and is therefore more difficult to design properly.

The influence of finished materials on life safety is reflected in such fire characteristics as combustibility, the rate of flame spread, smoke production, toxicity of gases and the rate of heat release. An ideal interior finish including furnishings and decorations, would be noncombustible. In reality the surfaces of walls, ceilings and floor coverings are usually made of combustible materials. Life safety provisions should then involve restrictions on smoke-producing capability and flame propagation. Since most of the usual plain materials for interior finish would not qualify, coating with a flame retardant may be necessary.

All building service equipment and utilities should be designed, installed and operated in accordance with rational engineering practices. For example, when designing exhaust air systems, public corridors must not be used as part of the air supply and return for obvious reasons. The corridors and paths to exits are usually pressurized to prevent the movement of smoke into these vital evacuation areas.

Building services and utilities should include appropriate fire protection equipment and automatic fire alarm systems. They must be periodically properly maintained and tested. Of course, they cannot, and must not, be a substitute for architectural protection measures. However, some of the restrictions placed on the requirements for life safety in the building concerning the number and size of exits and the type and characteristics of interior finish may be eased in sprinklered buildings.

Automatic sprinklers also serve as automatic fire alarms. However, it is better to have automatic fire detection and alarm systems in all buildings, be they sprinklered or unsprinklered, since they are vital in notifying the occupants of a fire. There is no substitute for an early warning of fire as the most important means of fire protection in buildings. Late notification of fire is one of the main causes of fire victims (Malhotra 1984). The other ones are smoke in evacuation paths, ignorance of alternative means of egress, improper design of the length and capacity of exits and obstructions on exits and exit doors (Malhotra 1984). A good review of the existing concepts for life safety analysis has been given by Nelson (1987).

8

Legislative control

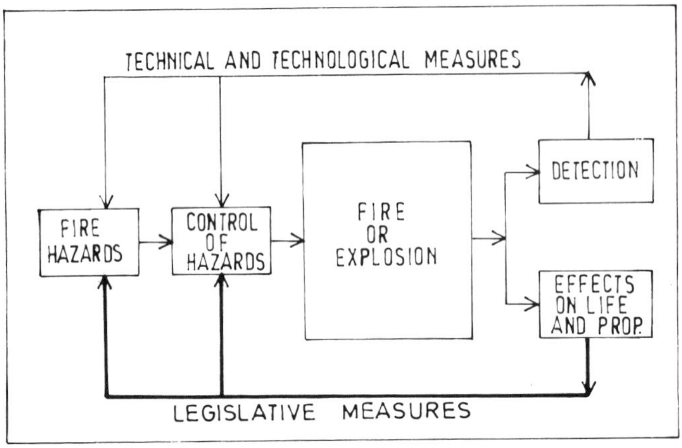

The effects of fires and explosions on life and property make the rationale for legislative control of sources of hazard. The aim of legislative measures is to control sources of hazards to such an extent that fires and explosions are prevented, or if they do occur, that no significant effects on people and the environment follow. Ideally, legislative measures, realized in the form of laws, codes, standards, instructions, etc., should act in harmony with technical and technological measures to achieve the optimum possible protection.

Historically, fire and explosion protection has always been an activity of special national interest. The importance of saving people's lives and property is reflected in passing mandatory administrative measures for fire and explosion control. In contrast to technical and technological measures, legislative measures are usually fairly rigid and more difficult to keep updated.

In a broader sense, legislative control also encompasses measures such as organizational, educational and propagandistic. Organizational measures involve all measures and activities intended to secure a normal working activity in hazardous or potentially hazardous conditions, that is to ensure both prevention and protection against fires and explosions. The prerequisite for the implementation and realization of organizational measures is the existence of qualified, educated people, the provision of financial sources and the provision of fire-fighting equipment.

Educational fire prevention and protection measures are designed to enhance the firesafety awareness of people and their ability to identify and correct specific fire problems. Similar objectives can also be obtained by fire and explosion prevention propaganda. These measures are necessary because experience reveals that most people lack the knowledge of what to do in the event of fire. The education, planning of emergency procedures and conducting of drills can remedy this problem.

There is a widespread feeling among the public that numerous technical and technological measures to control fire and explosion hazards, especially in industry, would not be voluntarily imposed if there were no legislative measures to enforce them. While this may be true for some sectors of industrial and public enterprise, the fact is that no reponsible company or party could afford to be without the self-imposed managerial hazard control since the best deterrent of indifference in protective measures is potentially huge financial losses.

A further hindrance to a more positive attitude toward both passive and active fire and explosion protection measures may be the apparent remoteness of such events in many practical situations. Many people have never experienced a fire or explosion at their work or home and tend to regard legislative measures as a nuisance. Hence violations of the provisions of fire and explosion legislation are common. In an analysis of the results of inspections of various premises it was found in *ca*. 65% cases that fire and explosion protection measures required by law were not implemented (Griffiths 1986). The most frequent contraventions were those directly concerning life safety (for example, the lack of secured evacuation routes, the absence of or nonfunctioning of fire-resistive doors, the lack of or inoperative automatic detection and alarm systems, inadequate illumination of the means of egress, improper marking of the means of egress and indications of directions, and the lack of fire extinguishers (Griffiths 1986, *Fire* 1986a).

Legislative measures for the control of fire and explosion hazards could be structured in the way Marshall discussed for the legislative control of major chemical hazards (Marshall 1984). Thus, the broad strategic measures could be identified with the supra-national control, while the tactical measures could correspond to control by plant management. Unfortunately little has been achieved towards the unification of fire and explosion legislative control on a supra-national level, although at many international meetings, calls for common standards have been voiced (*Fire Int*. 1986, Coster 1985, *BRE News Fire Res*. 1987). Because of the lack of international coordination there is unnecessary duplication of work on many subjects concerning fire and explosion protection. For example, in Europe alone there are at least five international organizations working independently on fire detection (Coster 1985). Yet, the trends toward regional and global associations will undoubtedly necessitate supra-national legislative control.

On the other hand, on national levels there is a tendency to overregulation, which might have originated owing to the highly interdisciplinary nature of fire and explosion protection. Warnings such as 'sea of standards' have been rightly expressed (*Fire* 1986b). Many technical disciplines have their own standards which taken together make coordinated, well-balanced fire and explosion legislative control very difficult.

References

Ahonen, A., Weckman, H., & Yli-Pentilla, M. (1985) Application of oxygen-consumption calorimetry to non-combustibility testing. *Fire and Materials* **9** No. 3, 135–144.

Altman, G. & Pfeiffer, R. (1984) Transformatoren in elektrischen anlagen aus der sicht des brandschutzes. *Elektrotechnik (Schweiz.)* **35** No. 12, 31–35.

Anderson, J. R. (1986a) Water supply requirements for fire protection. In: Cote, A. E. & Linville, J. L. (eds) *Fire Protection Handbook*, 16th edn., National Fire Protection Association, Quincy, MA, pp. 17–33 to 17–38.

Anderson, J. R. (1986b) Water distribution systems. In: Cote, A. A. & Linville, J. L. (eds) *Fire Protection Handbook*, 16th edn, National Fire Protection Association, Quincy, MA, pp. 17–39 to 17–59.

Anthony, E. J. (1977/78) The use of venting formulae in the design and protection of building and industrial plant from damage by gas or vapor explosions. *J. Hazar. Mater.* **2** 23–49.

ASHRAE (1983) Design of smoke control systems for buildings. American Society of Heating, Refrigerating and Air Conditioning Engineers, Atlanta, GA.

Babrauskas, V. (1984) Development of the cone calorimeter — a bench-scale heat release rate apparatus based on oxygen consumption. *Fire and Materials* **8** 81–95.

Babrauskas, V. (1986a) Comparative rates of heat release from five different types of test apparatuses. *J. Fire Sci.* **4** No. 2, 148–159.

Babrauskas, V. (1986b) Tables and charts. In: Cote, A. E. & Linville, J. L. (eds) *Fire Protection Handbook*, 16th edn, National Fire Protection Association, Quincy, pp. 5–125..

Babrauskas, V. (1986c) Free burning fires. *Fire Safety J.* **11**, No. 1–2, 33–51..

Baker, W. E. (1973) *Explosions in Air*. University of Texas Press, Austin & London.

Baker, W. E., Cox, P. A., Westine, P. S., Kulesz, J. J., & Strehlow, R. A. (1983) *Explosion Hazards and Evaluation*, Elsevier, Amsterdam.

Bamert, A. E. (1985) The importance of structural fire prevention measures. In: *Fires Build.* Proc. Eur. Symp., Luxembourg, 18–21 Sept., 1984, London, pp. 69–85.

Banks, J. & Montgomery, D. C., (1983) The analysis of three natural experiments relating to fire death rate. *J. Fire Sci.* **1** 33–35.

Bardon, M. F. & Fletcher, D. E. (1983) Dust explosions. *Sci. Prog. Oxf.* **68** 459–473.

Bartnecht, W. (1981) Explosions — Course, Prevention, Protection. Springer-Verlag, New York.

Beech, P. A. (1983) The effects of smoke and toxic gases. *Fire Eng. J.* **43** No. 131, 40–41..

Benjamin, J. A. (1984) The challenge of smoke. *Fire Safety J.* **7** 3–9.

Braun, E. & Levin, B. C. (1986) Polyesters: a review of the literature on products of combustion and toxicity. *Fire and Materials* **10** 107–123.

BRE News Fire Res. (1987) May, 3.

Brown, S. K. & Martin, K. G. (1985) Toxic gas and smoke measurement with the British Fire Propagation Test. II. Hydrogen chloride evaluation from UPVC building product. *Fire and Materials* **9** 95–102.

Brown, S. K. & Martin, K. G. (1986) Fire and building materials: measurement of smoke and toxic gas emission. *Fire J.* **11** No. 3, 16–18.

Bryan, J. L. (1986) Human behavior and fire. In: Cote, A. E. & Linville, J. L. (eds) *Fire Protection Handbook*, 16th edn, National Fire Protection Association, Quincy, MA, pp. 1–198 to 1–31.

BS 5306 — Fire extinguishing installations and equipment on premises, British Standards Institution, London.

Budnick, E. K. (1984) Estimating effectiveness of state-of-the-art detectors and automatic sprinklers on life safety in residential occupancies. *Fire Technol.* **20** No. 3, 5–22.

Bull, J. P. & Lawrence, J. C. (1979) Thermal conditions to produce skin burns. *Fire and Materials* **3** No. 2, 100–105.

Burry, P. (1985) Optical smoke detectors. *Fire Surv.* **14** No. 6, 18–22.

Butcher, E. G., Chitty, T. B., & Ashton, L. A. (1966) Temperatures attained by steel in building fires. JFRO, Fire Research Technical Paper No. 15.

Butcher, E. G. & Parnell, A. C. (1979) *Smoke Control in Fire Safety Design*. E. & F. N. Spon, London.

Butcher, E. G. & Parnell, A. C. (1985) Analysis of fire behavior. *Fire Surv.* **14** No. 3, 27–32.

Butcher, E. G. & Parnell, A. C. (1983) *Smoke Control in Fire Safety Design*. Russian translation of the original, Stroiizdat, Moskva, p. 18.

Cagliostro, D. E. & Islas, A. (1982) The effects of reduced oxygen and carbon monoxide on performance in a mouse pole-jump apparatus. *J. Combust. Toxicol.* **9** 187–193.

Campbell, J. A. (1986) Confinement of fire in buildings. In: Cote, A. E. & Linville, J. L. (eds) *Fire Protection Handbook*, 16th edn, National Fire Protection Association, Quincy, MA, pp. 7–109 to 7–124.

Charbonnier, J. (1984) Risques et disestres. *Rev. Secur.* **20** No. 208, 29–35.

Christian, S. D. & Nolan, P. F. (1985) Performance and relevance of smoke tests for material selection for improved safety in fires. *Flame Retard. '85.*, 2nd Int. Conf., London, 28–29 Nov., 1985, Luton, s.a., 17/1–17/10.

Clark, W. E. (1984) Sudden death: how firefighters get killed. *Fire Chief* **28** No. 10, 38–40.

Clarke, B. F. (1983) Control of materials based on smoke toxicity. *Int. Fire Chief* **49** No. 7, 24–25.

Connelly, E. M. & Harvey, C. S. (1986) System concepts for building firesafety. In: Cote, A. E. & Linville, J. L. (eds) *Fire Protection Handbook*, 16th edn, National Fire Protection Association, Quincy, MA, pp. 7–15.

Connor, S. (1984) Carbon monoxide — the unknown gas. *Fire Int.* **8** No. 87, 41–43.

Corbert, G. P. (1985) Dust explosion destroys New Jersey flour mill. *Fire Eng. J.* **45** No. 138, 22–24, 26.

Coster, G. D. (1985) Fire detection systems. In: *Fires Build., Proc. Eur. Symp., Luxembourg*, 18–21 Sept., 1984, London, pp. 369–373.

Cote, A. E. (1986) Assessing the magnitude of the fire problem. In: Cote, A. E. & Linville, J. L. (eds) *Fire Protection Handbook*, 16th edn, National Fire Protection Association, Quincy, MA, p. 1–2.

Cotton, F. A. & Wilkinson, G. (1972) *Advanced Inorganic Chemistry: a Comprehensive Text.* Interscience, New York, p. 354..

Cox, A. P. & Pasman, H. J. (1983) Risk analysis in the process industries. *Inst. Chem. Eng. Symp. Ser.* No. 80, G2–G7.

Cruice, W. J. (1986) Explosions. In: Cote, A. E. & Linville, J. L. (eds) *Fire Protection Handbook*, 16th edn, National Fire Protection Association, Quincy, MA, pp. 4–17 to 4–23.

Crummett, W. B. & Townsend, D. I. (1984) The trace chemistry of fire hypothesis: review and update. *Chemosphere* **13** No. 7, 777–788.

DeHaan, N. (1986) Interior finish. In: Cote, A. E. & Linville, J. L. (eds) *Fire Protection Handbook*, 16th edn, National Protection Association, Quincy, MA, pp. 7–74.

Drinkwater, A. E. (1985) Automatic fire-fighting systems: gaseous and powder systems. In: *Fires Build*, Proc. Eur. Symp., Luxembourg, 18–21 Sept., 1984. London, pp. 103–109.

Drysdale, D. P. (1985) *An Introduction to Fire Dynamics*. Wiley, Chichester.

Drysdale, D. P. & Abdul-Rahim, F. F. (1985) Smoke production in fires: small-scale experiments. Fire Safety: Sci. and Eng. Symp. Denver, Colo., 26–27 June, 1984. Philadelphia, Pa, 285–300.

Durrant, C. (1985) Time to formulate revised role for fire defence within UK industry. *Fire* **77** No. 955, 34–35.

Egan, M. R. & Litton, C. D. (1986) Wood crib fires in a ventilated tunnel. Rept. Invest. Bur. Mines. US Dep. Inter., No. 9045..

Einhorn, I. N. (1983) Toxicity of combustion products. *Int. Fire Chief* **49** No. 7, 13–15.

Ellis, W. P. & Biegel, S. L. (1982) State-of-the-art toxicity effects resulting from fires in buildings. *J. Comb. Toxicol.* **9** 139–162.

Emmons, H. W. (1978) The prediction of fires in buildings. In: *Proceedings of the Seventeenth International Symposium on Combustion*, The Combustion Institute, Pittsburg, PA, 1978, pp. 1101–1111..

Emmons, H. W. (1981) The calculation of fire in a large building. ASME 20th Meeting Nat. Heat Transfer Conf. Milwaukee.

Emmons, H. W. & Atrega, A. (1982) The science of wood combustion, *Indian Acad. Sci. Eng. Sci.* **5** 259–268.

EN 54 (1976) European standard "Components of automatic fire detection systems — Part 1", European Committee for Standardization, Brussels.

Fardell, P. J., Murrell, J. M., & Murrell, J. V. (1986) Chemical "fingerprint" studies of fire atmospheres. *Fire and Materials* **10** 21–28.

Feek, G., Baptiste, M. S., & Tate, C. L. (1979) Burn injuries: epidemiology and prevention. *Accid. Anal. Prev.* **11** No. 2, 129–136.

Field, P. (1983) Explosibility assessment of industrial powders and dusts. Building Research Establishment Report, Her Majesty's Stationery Office, London.

Fire (1985) More die from smoke or gas than from burns in fires. **77** No. 956, 8.

Fire (1986a) Heavy fines for DIY company on FP Act contraventions. **79** No. 977, 6.

Fire (1986b) Warning about "sea of standards". **79** No. 975, 6.

Fire (1986c) **79**, No. 976, 21.

Fire Chief (1985) Smoke researcher pinpoints lethal agents. **29** No. 3, 12–13.

Fire Contr. Dig. (1984a) World leading fire death rate blamed on American indifference. **10** 3–4.

Fire Contr. Dig. (1984b) 1983 fire-fighting deaths figures released by NFPA. **9** No. 7, 9.

Fire Int. (1983) The World Fire Statistics Centre. **7** No. 28, 45–47.

Fire Int. (1986) Triple call for common standard. **10** No. 100, 74–76.

Fire J. (1984) Toxicity of smoke. **9** No. 4, 28–29.

Fir. J. (1985) "Light water" AFFF/alcohol type concentrate: the most versatile firefighting foam. **79** No. 1, 1.

Fire Prev. (1983) Multiple-death fires in the United States. No. 163, 7.

Fire Prev. (1984) The law on fire precautions in buildings. Part 3: Factories. No. 166, 22–27.

Fire Prev (1986 No. 189, 41.

Fire Prot. (1985a) Protecting against the false alarm. **48** No. 579, 14–15.

Fire Prot. (1985b) Fire losses still at high level. **48** No. 575, 3.

Fire Protection Handbook (1977) 14 edn, National Fire Protection Association, Boston, MA, p. 1–14.

Fitzgerald, R. W. (1986) Fundamentals of firesafe building design. In: Cote, A. E. & Linville, J. L. (eds) *Fire Protection Handbook*, 16th edn, National Fire Protection Association, Quincy, MA, pp. 7–2 to 7–9.

Fowell, A. J. (1984) An approach to hazard assessment of combustion products in building fires. Proceedings of the Conference "Flame retardancy advances in fire safety: regulations, testing, products, markets". Callaway Gardens, Pine Mountain. Georgia, 1984, pp. 24–35.

Gad, S. C. & Smith, A. C. (1983) Influence of heating rates on the toxicity of evolved combustion products: results and a system for research. *J. Fire Sci.* **1** 465–479.

Gormsen, H., Jeppesen, N., & Lund, A. (1984) The causes of deaths in fire victims. *Forensic Sci. Int.* **24** 107–111.

GOST 12.1.017-80 (1980) Pozarovzrivoopasnost nefteproduktov i himiceskih organiceskih produktov, Moskva.

Griffiths, A. (1986) Full implications of improved notices must be considered. *Fire* **79** No. 977, 8.

Grimwood, P. T. (1985) Getting the best possible results from ventilation. *Fire* **77** No. 64, 33–34, 36–38.

Gross, D. & Robertson, A. F. (1965) Experimental fires in enclosures. Tenth International Symposium on Combustion, The Combustion Institute, p. 931.

Haessler, W. M. (1974) Extinguishment of fire. National Fire Protection Association, Quincy, MA.

Haessler, W. M. (1986) Theory of fire and explosion control. In: Cote, A. E. & Linville, J. L. (eds) *Fire Protection Handbook*, 16th edn, National Fire Protection Association, Quincy, MA, p. 4–43.

Hagglund, A. (1983) Deaths from fire in sprinkler-protected buildings. *Fire Technol.* **19** 210–214.

Hagglund, B. (1984) A room fire simulation model. *Fire and Materials* **8** 105–111.

Hagglund, B., Jansson, R. & Onnermark, B. (1974) Fire development in residential rooms after ignition from nuclear explosions. FOA Rep. C 20016-D6 (A3), Forsvarets Forskningsanstalt, Stockholm, Sweden.

Hall, J. R. (1984) When fire fighters die in groups — patterns and scenarios. *Fire Command* **51** 32–33.

Harikae, H. & Morita, I. (1984) Measurement of firefighter air consumption during fire operations. *Rept. Fire Sci. Lab.* No. 21, 133–139.

Harmathy, T. Z. (1972) A new look at compartment fires, parts I and II. *Fire Technol.* **8** Nos. 3 & 4, 196 & 326.

Harmathy, T. Z. (1976) Design of buildings for fire safety — Part I. *Fire Technol.* **12** Nos. 2 & 3, 95 & 219.

Harmathy, T. Z. (1983) Overview of impact of energy conservation on firesafety. *Fire Technol.* **19** 31–56.

Harmathy, T. Z. (1984) Burning, pyrolysis, combustion and char oxidation. Need for clarifying terminology. *Fire and Materials* **8** 224–226.

Harmathy, T. Z. (1985) Thermal inertia or thermal absorptivity? *Fire Technol.* **21** 146–149.

Harmathy, T. Z. (1986) Post-flashover fires — an overview of the research at the National Research Council of Canada (NRCC). *Fire Technol.* **22** 210–233.

Harmathy, T. Z. & Mehaffey, J. R. (1983) Post-flashover compartment fires. *Fire and Materials* **7** No. 2, 49–61.

Harris, R. J. (1983) *Gas Explosions in Buildings and Heating Plants*. E. & F. N. Spon, London.

Hasemi, J. & Shigekawa, K. (1981) Radiation tolerance criterion of people in big fires. *Bull. Jap. Assoc. Fire Sci. Eng.* **31**, No. 1, 7–14.

Heibrock, W. & Weis, A. (1983) Automatische brandmeldeanlagen — aufbau und funktion, planungsfrundsatze, hineweise, bewertung durch die sachversicherer. *Maschinenshaden* **56** No. 6, 232–241.

Helsper, C., Fissan, H., Muggli, J. & Scheidweiler, A. (1983) Verification of ionization chamber theory. *Fire Technol.* **19** No. 1, 14–21.

Hertel, H. (1985) Neue Prüfgrundsätze für schwerentflammbare und nichtbrennbare baustoffe. *Bauphysik* **7** No. 4, 112–118.

Hildebrand, J. (1984) Stress research: part 5. Availability of services. *Fire Command* **51** No. 11, 32–35.

Hinderer, R. K. (1984) A comparative review of the combustion toxicity of polyvinyl chloride. *J. Fire Sci.* **2** 82–97.

Hodnett, R. M. (1986a) Water and water additives for fire fighting. In: Cote, E. A. &

Linville, J. L. (eds) *Fire Protection Handbook*, 16th edn, National Fire Protection Association, Quincy, MA, pp. 17–2 to 17–11.

Hodnett, R. M. (1986b) Automatic sprinkler systems. In: Cote, A. E. & Linville, J. L. (eds) *Fire Protection Handbook*, 16th edn, National Fire Protection Association, Quincy, MA, pp. 18–3 to 18–26.

Hodnett, R. M. (1986c) Water spray protection. In: Cote, A. E. & Linville, J. L. (eds) *Fire Protection Handbook*, 16th edn, National Fire Protection Association, Quincy, MA, pp. 18–77 to 18–82.

Hoffmann, W. (1983) Die schwerpunkte für einen wirksamen brandschutz. *Textilbetrieb* **101** No. 12, 34.

Huggett, C. (1980) Estimation of rate of heat release by means of oxygen consumption measurements. *Fire and Materials* **4**, 226.

Huggett, C. (1984) Combustion conditions and exposure conditions for combustion product toxicity testing. *J. Fire Sc.* **2** 328–347.

Humphreys, H. & Christopher, A. J. (1985) Acidic volatiles from overheated polymeric materials. *Fire and Materials* **9** 120–124.

Hurst, N. W. & Jones, T. A. (1985) Review of products evolved from heated coal, wood and PVC. *Fire and Materials* **9** 1–8.

Ingberg, S. H. (1928) Tests of the severity of building fires. *NFPA Quarterly* **42** 43.

Int. Fire Chief (1986) FTC rules against vanguard heat detectors. **52** No. 6, 6–7.

Int. Fire Chief (1987) Fire measurement tool. **53** No. 3, 6.

ISO 1182 (1983) Fire tests — building materials — noncombustibility test, 2nd edn, International Organization for Standardization, Genève.

Kaiser, J. (1979/80) Experience of the Gretener method. *Fire Safety J.* **2** 96.

Kaufman, C. W., Srinath, S. R., & Tai, C. S. (1987) *Proceedings of the International Symposium on the Explosion Hazard Classification of Vapors, Gases and Dusts*, Washington DC, 15–18 July 1986. Washington, DC, Nat. Acad. Press, 1987, VIII, pp. 41–46.

Kawagoe, K. (1958) Fire behavior in rooms. Building Research Institute, Japan, Report No. 27.

Kawagoe, K. (1967) Estimation of fire temperature–time curve in rooms. Building Research Institute, Japan, Research Paper No. 29..

Keith, M. (1985) The use of water spray barriers to disperse spills of heavy gases. *Plant/Oper. Prog.* **4** No. 4, 234–241.

Kimstach, I. F. , Devlishchev, N. M., & Evtyushchkin, N. M. (1984) *Fire Tactics*. Stroiizdat, Moscow.

Kirby, B. R. (1986) Recent developments and applications in structural fire engineering design — a review. *Fire Safety J.* **11** No. 3, 141–151, 169–179.

Krasny, J. F. & Sello, B. (1986) Fibers and textiles. In: Cote, A. E. & Linville, J. L. (eds), *Fire Protection Handbook*, 16th end, National Fire Protection Association, Quincy, MA, 5–17.

Kunzi, H. (1980) Routine investigations of thermic hazards: some uneasiness remains with the examiner. *3rd Int. Symp. Loss Prevent. and Safety Promot. Process Ind.*, Basel, Sept. 1980, pp. 5/386–5/394.

Law, M. (1963) Heat radiation from fires and building separation. Technical paper No. 5, Joint Fire Research Organization. Borehamwood, England.

Lengauer, K. (1986) Strahlengefahren durch ionisationsrauchmelder? *Brennpunkt* **38** No. 2, 20.

Levin, B. C., Paabo, M., Fultz, M. L., & Bailey, C. S. (1985) Generation of hydrogen cyanide from flexible polyurethane foam decomposed under different combustion conditions. *Fire and Materials,* No. 3, 125–134.

Lie, T. T. (1974) Characteristic temperature curves for various fire severities. *Fire Technol.* **10** 315–326.

Longwell, J. P. (1982) The formation of polycyclic aromatic hydrocarbons by combustion. In: *Proceedings of the 19th International Symp. on Combustion*, Haifa, 8–13 August, Pittsburg, PA, pp. 1339–1394..

Lowe, D. R. T. & Solomon, C. H. (1983) Hazard identification procedures. *Inst. Chem. Eng. Symp. Ser.* No. 80, pp. G8–G24.

Macdonald, J. N. (1985) Non-evacuation in compartmented fire resistive buildings can save lives and it makes sense. US Dep. Commer. Nat. Bur. Stand. Spec. Publ. No. 694, 117–130.

Magnusson, S. E. & Thelandersson, S. (1970) Temperature–time curves of complete process of fire development. Theoretical study of wood fuel fibres in enclosed places. *Acta Polytechnica Scandinavica*, Civil Engineering & Building Construction Series No. 65, Stockholm.

Maguire, H. (1984) Medical examiner's protocol: a step in the right direction. *Fire Chief* **28** 57–59.

Malhotra, H. L. (1984) Escape from fire. In: *Fires Build.*, Proc. Eur. Symp., Luxembourg, 18–21 Sept., 1984, London, pp. 115–125, 463–465, 564–566.

Marshall, V. C. (1980) Historical and theoretical approaches to the prediction of hazard and risk. *3rd Int. Symp. Loss Prevent. and Safety Promot. Process Ind.*, Basel, Sept. 1980, pp. 6/395–6/408.

Marshall, V. C. (1984) Shaping a strategy of control. *The Chemical Engineer*, June 1984, 447.

Marshall, V. C . (1987) *Major Chemical Hazards*. Ellis Horwood, Chichester.

McCaffrey, B. J., Quintiere, J. G., & Harkleroad, M. F. (1980) Estimating room temperatures and the likelihood of flashover using fire test data correlations. National Bureau of Standards, Washington, DC.

McGarry,. F. (1983) Toxicity background factors and the New York State study. *Int. Fire Chief.* **49** 19–21.

McGuire, A. (1983) Cigarettes and fire deaths. *N.Y. State J. Med.* **83** 1296–1298.

Mehaffey, J. R. & Harmathy, T. Z. (1981) Assessment of fire resistance requirements. *Fire Technol.* **17** 221.

Middleton, J. F. (1983) Developments in flame detectors. *Fire Safety J.*, No. 3, 175–182.

Miyama, J. & Saito, F. (1982) Fire detection and smoke properties. US Dept. Commer. Nat. Bur. Stand. Spec. Publ., No. 639, 31–38..

Moore, D. W. (1986) Halogenated agents and systems. In: Cote, E. A. & Linville, J. L. (eds) *Fire Protection Handbook*, 16th edn, National Fire Protection Association, Quincy, MA, pp. 19–11 to 19–23.

Müller, H. G. (1983) Von der flamme zum flammenmeer Istanbul hat viele parallelen. *Brennpunkt* **35** 4–5.

Nelson, H. E. (1987) Concepts for life safety analysis. *Fire Safety J.* **12** No. 3, 167–177.

Newman, B. G. (1982) The hazard of taking the heat. *Fire Sci. Today* **49** No. 8, 22–23.

NFPA 10 (1984) Standard for portable fire extinguishers. National Fire Protection Association, Quincy, MA.

NFPA 12 (1985) Standard on carbon dioxide extinguishing systems. National Fire Protection Association, Quincy, MA.

NFPA 49 (1975) Hazardous chemicals data. National Fire Protection Association, Quincy, MA.

NFPA 69 (1978) Standard on explosion prevention systems. National Fire Protection Association, Quincy, MA.

NFPA 70 (1984) National electric code. National Fire Protection Association, Quincy, MA.

NFPA 71 (1985) Standard for the installation, maintenance, and use of central station signaling systems. National Fire Protection Association, Quincy, MA.

NFPA 72A (1985) Standard for the installation, maintenance and use of local protective signaling systems for guard's tour, fire alarm and supervisory service. National Fire Protection Association, Quincy, MA.

NFPA 72B (1986) Standard for the installation, maintenance and use of auxiliary protective signaling systems. National Fire Protection Association, Quincy, MA.

NFPA 72C (1986) Standard for the installation, maintenance and use of remote station protective signaling systems. National Fire Protection Association, Quincy, MA.

NFPA 73D (1986) Standard for the installation, maintenance and use of proprietary protective signaling systems. National Fire Protection Association, Quincy, MA.

NFPA 72E (1984) Standard on automatic fire detectors. National Fire Protection Association, Quincy, MA.

NFPA 77 (1983) Recommended practice on static electricity. National Fire Protection Association, Quincy, MA.

NFPA 78 (1983) Lightning protection code. National Fire Protection Association, Quincy, MA.

NFPA 80A (1980) Recommended practice for protection of buildings from exterior fire exposures. National Fire Protection Association, Quincy, MA.

NFPA 101 (1985) Code for safety to life from fire in buildings and structures. National Fire Protection Association, Quincy, MA.

NFPA 204M (1985) Guide for smoke and heat venting. National Fire Protection Association, Quincy, MA.

NFPA 550 (1986) Guide to the firesafety concepts tree. National Fire Protection Association, Quincy, MA.

NFPA 901 (1981) Uniform coding for fire protection. National Fire Protection Association, Quincy, MA.

Nodziri, T. & Marujama, K. (1984) Physiological changes of firefighter by weight of personal equipment under hot environment. *Rept. Fire Sci. Lab.* No. 21, 125–132.

Packham, S. C. (1984) Forensic applications of combustion toxicology. In: *Proceedings of the Workshop on Toxic Hazards from Fire*, Washington, DC, 15–16 Nov. 1984, Technomic, Lancaster, PA.

Palmer, K. N. (1973) *Dust Explosions and Fire*. Chapman and Hall, London.

Parikh, I. S. & Beyreis, I. R. (1985) Survey of the state of the art of mathematical fire modelling. US Dept. Commer. Nat. Bur. Stand. Spec. Publ. No. 694, 93–111.

Parker, W. J. & Long, M. E. (1972) Development of a heat release rate calorimeter at NBS. ASTM STP 502. American Society for Testing and Materials. Philadelphia, PA, 135–151.

Parry. L. A. (1983) Fighting a winning war against smoke. *Timber Trades J. and Wood Process.* **326** No. 5571, 22–23.

Paul, K. T. (1985) Relationship between smoke in large scale fire tests and in standard smoke test (XP2, NBS and modified Munich tests). *Flame Retard '85.*, 2nd Int. Conf., London, 28–29 Nov., 1985, Luton, s.a. 18/1–18/13.

Peissard, W. G. (1983) Der praktische were der automatischen brandentdeckung. Neue technologische trends in der relaisierung. *Textilbetrieb* **101** No. 12, 27–28, 31–32.

Peissard, W. G. (1985) Cerberus. Der $LiTaO_3$ — losung aus mannedorf. *Wirtschaftsschutz+Sicherheitstechn.* **7** No. 9, 370–375.

Peterson, P. (1986) Special systems for explosion damage control. In: Cote, E. A. & Linville, J. L. (eds) *Fire Protection Handbook*. 16th edn, National Fire Protection Association, Quincy, MA, pp. 12–75 to 12–81.

Philips, A. W., (1980) What should you know about burns? *Fire J.* **74** No. 1, 35–36.

Powell, F. (1986) Can non-sparking tools and materials prevent gas explosions? *Gaz–Eaux–Eaux usees* **66** No. 6, 419–428.

Queen, B. L. (1983) Fire dangers: the toxicity perspective. *ASTM Stand. News* **11** No. 12, 34–37.

Quintiere, J. G. (1977) Growth of fire in building compartments. In: Robertson, A. F. (ed), *Fire Standards and Safety*. ASTM STP 614, American Society for Testing and Materials, PA, pp. 131–167.

Quintiere, J. G. (1981) An approach to modelling wall fire spread in a room. *Fire Safety J.* **3** 201–214.

Quintiere, J. G. & Harkleroad, M. (1985) New concepts for measuring flame spread properties. In: Harmathy, T. Z. (ed) *Fire Safety: Science and Engineering*. ASTM STP 882. American Society for Testing and Materials, PA, 239–276.

Ramamchandran, G. (1985) The human aspects of fires in building — a review of research in the United Kingdom. In: Harmathy, T. Z. (ed) *Fire Safety: Science and Engineering*, ASTM STP 882, American Society for Testing and Materials, Philadelphia, pp. 386–422.

Ramachandran, G. & Chandler, S. E. (1984) Economic value of early detection of fires in industry and commercial premises, Building Research Esatablishment Information Paper IP 13/84, Borehamwood.

Rasbash, D. J. (1984) Recent developments in fire safety engineering. *Fire Surv.* **13** 16–27.

Rasbash, D. J., Drysdale, D. D., & Deepak, D. (1986) Critical heat mass transfer at pilot ignition and extinction of a material. *Fire Safety J.* **10** 1–10.

Reichwerger, Z. & Balbin, J. (1985) Towards regulating toxicity of materials in building. *Fire and Materials* **9** 117–119.

Rev. Techn. Feu (1979) Endurance à la chaleur. **20** No. 192, 38.

Rieber, M. (1983) Why it is difficult to measure the burning characterstics of materials and products. *Fire and Materials* **7** No. 4, 180–184.

Ris, J. (1985) Flammability testing state-of-art. *Fire and materials* **9** No. 2, 75–80.

Rockett, J. A. & Morita, M. (1985) The NBS/Hardward Mark VI multiroom fire simulation. *Fire Sci. Technol.* **5** 149–163.

Rohrbacher, H. (1983) Die Weiterentwicklung der automatichen brandmelder. *Wirtschaftsschutz+Sicherheitstechn.* **5** No. 5, 195–196, 198.

Rosander, M. & Giselsson, K. (1984) Making the best use of water for extinguishing purposes. *Fire* **77** No. 952, 43–46.

Ruegg, R. T. & Fuller, S. K. (1987) The economics of fast-response residential sprinkler systems. *Fire J.* **11** No. 2, 4–10, 21.

Runes, E. (1972) Explosion venting. In: *Loss Prevention,* Vol. 6, Proceedings of the 6th Symposium on Loss Prevention in the Chemical Industry, New York, pp. 63–67.

Schadenprisma (1983) Brandmeldeanlagen eine ubersicht uber die technische entwicklung. **12** No. 3, 41–44.

Schaenman, P. S. (1983) Why Europe isn't burning. *Fire Chief* **27** 43–48.

Seinfeld, J. H. (1975) *Air Pollution — Physical and Chemical Fundamentals.* McGraw-Hill, New York, pp. 1–4.

Sekine, H., Hodziri, T., & Sudzuki, J. (1982) Rept. Fire Sci. Lab. No. 19, 90–99.

Shaw, J. R. (1986) Fire retardant and flame resistant treatments of cellulosic materials. In: Cote, A. E. & Linville, J. L. (eds) *Fire Protection Handbook.* 16th edn, National Fire Protection Association, Quincy, MA, pp. 12–36.

Short, W. A. (1987) An overview of international, regional and national standards. *Proceedings of the International Symposium on the Explosion Hazard Classification of Vapors, Gases and Dusts*, Washington, DC, 15–18 July, 1986, Washington, DC, Nat Acad. Press, 1987, p. 343.

Simpson, A. R., Allen, N. L., & Dring, D. (1985) Lightning as a cause of fire and how to deal with it. *Fire Eng. J.* **45** No. 138, 30–31.

Smith, J., H. W. (1983) Wiederkehrende aspekte der sicherheitspolitik — in der unternehmensfuhrung. *Protector* **11** No. 5, 3–7.

Smith, S. R. (1985) Plastics — the hidden danger in our surroundings. *Fire Eng. J.* **45** No. 138, 22–24, 25–26.

Starrett, P. S. (1981) Factors influencing flame spread rates in solid materials. "Flammabil. Solid Plast. Pt. 2", Westport, Conn., 85–105.

Taylor, R. (1984) The real enemy is smoke. *ASHRAE J.* No. 4, 44–47.

Thomas, P. H. (1974) Fires in enclosures. Building Research Establishment, Current Paper, CP 30/74, Fire Research Station, Borehamwood.

Thomas, P. H. (1975) Factors affecting ignition of combustible materials and their behavior in fire. In: *Fire Safety of Combustible Materials.* Proceedings of International Symposium, University of Edinburgh, p. 84.

Thomas, P. H. (1983) Modelling of compartment fires. *Fire Safety J.* **5** 181–190.

Thomas, P. H. (1985) Fire development modelling. In *Proceedings of the European Symposium on Fires in Buildings*, Luxembourg, 18–21 Sept. 1984, London, New York, pp. 187–200.

Thomas, P. H., Haselden, A. J. M., & Law, M. (1967) Fully developed compartment fires: two kinds of behavior. JFRO, Fire Research Technical Paper No. 18.

Todd, C. S. (1987) Lack of maintenance jeopardises the benefits of active systems. *Fire Surv.* **16**, No. 5, 3, 6–11.

Tong, D. & Canter, D. (1985) The decision to evacuate: a study of the motivations which contribute to evacuation in the event of fire. *Fire Safety J.* **9** No. 3, 257–265.

Tuhtar, D. (1984) *Air and Water Pollution*, Svjetlost, Sarajevo, Yugoslavia, pp. 213–220 (in Serbocroatian).

Tyler, B. J. (1985) Using the Mond index to measure inherent hazards. *Plant/ Operations Progress* **4** No. 3, 172–175.

U.S. Environmental Protection Agency (EPA) (1970) Air Quality Criteria for Carbon Monoxide, AP-62. U.S. Govt. Printing Office, Washington, DC, pp. 8–18.

U.S. Environmental Protection Agency (EPA) (1982) Air quality criteria for oxides of nitrogen. Environmental Criteria and Assessment Office, Research Triangle Park, NC, EPA-600/8-82-026F, pp. 5–21.

Vandevelde, P. (1985) Testing and classification of the resistance to fire of building components. In: *Fires Build.*, Proc. Eur. Symp., Luxembourg, 18–21 Sept., 1984. London, pp. 201–212.

Walsh, H. (1986) A national mandate for smoke detectors. *Int. Fire Chief* **52** No. 9, 8–10.

Wardlaw, G. (1983) Behavior of people in emergencies. *Fire J.* **7** No. 4, 29–31.

Washborn, A. E., Harlow, D. W., & Fahy, R. F. (1986) US fire deaths 1985. *Fire Command* **53** No. 6, 20–27, 31, 34, 36–37.

Waterman, E. (1972) Room flashover model studies. *Fore Technol.* **8** 17.

Waterman, T. E. (1984) Fire venting of sprinklered buildings. *Fire J.* **78** No. 2, 30–32, 34, 36–39, 86.

Waters, D. (1985) Technique plays a major role in applying foam concentrates. *Fire* **78** No. 960, 10–12.

Wolf, P. C. (1971) *Environ. Sci. Technol.* **5** 213.

Wooley, W. D. (1985) Are foams a fire hazard? *Cellular Polymers* **4** 81–115.

Yao, C. (1973) Explosion venting of low strength equipment and structure. In: *Proceedings of the 8th Symposium on Loss Prevention in the Process Industries*, Vol. 8, Philadelphia, PA, 1.

Young, R. (1985) Why Europeans should invest more in sprinkler systems? *Fire* **77** No. 955, 17.

Zimmerman, C. E. (1986a) Automatic fire detectors. In: Cote, A. A. & Linville, J. L. (eds), *Fire Protection Handbook*. 16th edn, National Fire Protection Association, Quincy, MA, pp. 16–29 to 16–39.

Zimmerman, C. E. (1986b) Protective signalling systems. In: Cote, A. E. & Linville, J. L. (eds), *Fire Protection Handbook*. 16th edn, National Fire Protection Association, Quincy, MA, pp. 16–13 to 16–21.

Index